TALKING ROCKS

GEOLOGY AND 10,000 YEARS OF

NATIVE AMERICAN TRADITION IN THE

LAKE SUPERIOR REGION

Ron Morton and Carl Gawboy

Illustrations by Carl Gawboy

University of Minnesota Press

Minneapolis / London

Originally published by Pfeifer-Hamilton Publishers, 2000
First University of Minnesota Press edition, 2003

Published by the University of Minnesota Press
111 Third Avenue South, Suite 290
Minneapolis, MN 55401-2520
http://www.upress.umn.edu

ISBN 978-0-8166-4430-8 (pb)

A Cataloging-in-Publication record for this book is available
from the Library of Congress.

UMP LSI

*We dedicate this book to the many field geologists
and Native Americans, past and present, whose observations,
interpretations, and earth stories made this possible.*

CONTENTS

Acknowledgments

The beginnings of this book are to be found in a workshop designed and organized by Cindy Johnson-Groh, who was then the director of the Plus Center at The College of Saint Scholastica in Duluth, Minnesota. The workshop, "Reading the Land," was designed for teachers of grades four through eight to learn about the natural history and American Indian heritage of Minnesota. Part of the workshop involved a bus tour to explore Minnesota's ecology and geology and see how these could be integrated with Native American land ethics. The workshop was attended by thirty enthusiastic teachers, and it was their interest and questions, along with the lively lectures by Larry Weber, Cindy Johnson-Groh, and George Lokken, as well as the great bus tour, that turned an idea into a working collaboration.

For their help from the end of the workshop to the final version of this book, we express our appreciation to Tim Roufs, Charles Matsch, Penny Morton, and Cindy Donner, who graciously read early versions of the manuscript and/or individual chapters of the book. We are indebted to them for their helpful comments and numerous ideas for improvement.

In the end the final form of this book was the result of much prodding, intellectual pushing, and patience from Don Tubesing of Pfeifer-Hamilton. Without his help, ideas, and encouragement, this book would not have happened.

Author's Note

Carl and I first met at a workshop presented by the Plus Center at The College of Saint Scholastica in Duluth, Minnesota. The workshop, entitled "Reading the Land," was designed for fourth-through eighth-grade teachers to learn about the ecology and geology of Minnesota and how these could be integrated with Native American heritage and land perspectives.

During the workshop, listening to Carl talk about Native Americans who have lived in the midcontinent region for more than 10,000 years, I was struck by how close these people were to planet earth and how much of their culture was related to, and interwoven with, geology. I was also impressed with the way Carl blended the facts of Native American culture and tradition with different stories and myths. Carl's lectures started me thinking about this book, and the more I considered it, the more excited I became.

At supper one evening, in a kind of garbled rush, I outlined my ideas to Carl and was amazed to find he had been thinking along much the same lines. According to Carl, many of the geological topics presented at the workshop could be found in Native American stories and traditions. Geology, Carl said, provided a foundation for many of the things Native Americans had observed and believed for more than 10,000 years.

As the workshop ended, our collaboration began. We started to write a book that integrated geology and living with planet earth, as it related to the culture, science, and heritage of Native Americans living in the midcontinent region. However, as we soon discovered, in order to do justice to such a project we had to go beyond science, myths, and traditions. The book also had to be about the search for common ground between two different ways of seeing planet earth and the land we live on and need.

In our explorations of the midcontinent region, Carl and I came to the conclusion that not only did these different worldviews, the Western view of planet earth as seen through the eyes of geology and the more holistic view of the land and planet as seen through the eyes of Native Americans, share a lot of common ground, but each actually strengthened and gave balance to the other.

In the end, then, *Talking Rocks* turned out to be not only a story of geology and how this geology affected and influenced the traditions, myths, and science of Native Americans for more than 10,000 years, but also a story of two people, from culturally different backgrounds, trying to understand and appreciate different ways of seeing this land and planet we call home.

The other difficult aspect of writing this book turned out to be finding a suitable and entertaining way to present and link together the different stories we wanted to include in the book. We struggled with this problem for some time without finding a satisfactory solution. It was Carl, or rather Carl in action, who finally provided the answer.

I was previewing slides for an Elderhostel course I was to teach on the Gunflint Trail when I came across one of Carl. Taken on a spring day at Jay Cooke State Park, it showed him demonstrating how the voyageurs transported goods and furs across portages. The very next slide caught him in action at Rice Lake. Under a deep blue sky Carl, in a canoe, demonstrated how Native Americans harvested wild rice.

These two slides, the memories that went with them, and the lectures Carl and I had presented together gave form and substance to both the Native American storyteller and philosopher found throughout the pages of this book and the geologist who narrates this story. Together, they allowed us to interweave two different worldviews and the search for common ground with the stories we wanted to tell. At the same time these two characters give full voice to much that is within and between

Carl and me. Finally, though the geologist and the Native American we call Earth Walks are fictional characters, the geological and cultural descriptions and events in this book are as factual and accurate as our understanding of geology, astronomy, and history can make them.

Ron Morton
1999

CHAPTER ONE

THE MEETING

Pouring myself a glass of seventeen-year-old Bowmore, I sat back in a soft, wide chair and thought about everything that had occurred over the past few months. I thought especially about small things and how these could turn your world upside down, or downside up, as an 8-year-old girl once told me.

A single quote: such a small thing, and yet, because of it, I would never again see my profession—heck, why not admit it, my addiction—in exactly the same way. Taking a sip of scotch, I had to smile as I remembered the first geology course I had ever taken. It was during my first year at college, and I fell completely under the spell of volcanoes, dinosaurs, glacial ice, and the moving continents of planet earth. It turned out this was love at first sight; it was then and there I decided I wanted to be a geologist.

Ever since that first geology course I have seen rocks and minerals as beautiful and mysterious things. They have been my sunsets, rainbows, and Camelots rolled into one. I honestly believed planet earth had placed them beneath my feet so I could solve the mystery of who and what they were, discover how they came into being, and unearth any treasures they might hide. Rocks were magic carpets, and through them I could travel over more than 3 billion years of earth history to walk on landscapes as bare and alien as those of Mars and Venus, stroll across wrinkled fields of lava picking up agates and nuggets of copper as I went, or ride atop massive ice sheets as they dug great lakes and devoured the land.

Rocks and minerals and the processes that created them were the essence of planet earth, and in understanding them I believed I came closer to understanding myself.

Just as importantly, I saw geology as the foundation upon which modern civilization stood. Without geology many of the things we take for granted would never have been built, assembled, or invented. Computers, airplanes, and pencils; cars, microwave ovens, satellite television, and a million other necessities, gadgets, and devices—all testament to the truth "if you don't grow it, you mine it," and if you mine it, it's because of geology and geologists.

That's the way it was until fate came knocking on my door. Fate turned out to be a Native American named Earth Walks.

Earth Walks, during our first few meetings, was more than happy to inform me there was only one thing wrong with my view of planet earth: it had no balance. According to him I looked at planet earth from one and only one perspective. I saw with only one eye. It was like looking at a rainbow and seeing only one color.

Earth Walks, at the time, seemed just as one-eyed, though his perspective and mine were poles apart. He appeared to be a traditionalist with a capital T, believing in and practicing many of the old ways, especially those that pertain to our relationship with planet earth. It was only later, after we got to know each other, that I found him to be much more complex than this.

Overall, modern science and Earth Walks do not see eye-to-eye. He once told me Western science was like Frankenstein's monster; it had no soul, and without a soul there could be no respect. In his view science, as far as land use and management of natural resources go, has caused more problems than it has ever solved, and most of these are due to the Western-based philosophy of control over, and exploitation of, planet earth. For him the answers to our land and resource problems are to be found in the wisdom of Native Americans, a people who, he claims, possess intimate and detailed knowledge of planet earth, knowledge derived from thousands of years of direct contact and careful observation.

Earth Walks believes we have a sacred responsibility to the planet and to all creatures that live upon her. We are all part of the circle, all part of what he calls Mother Earth, and to believe we can control and exploit her as we please is ignorance of the darkest kind.

Fate and I met about a year ago. I had been invited to give a series of lectures for nonscientists on the geology of the Great Lakes region. These geological lectures were to be integrated with lectures on early Native Americans, emphasizing how these people lived with and adapted to the changing land over the last 10,000 years. The organizers of the lecture series arranged a meeting between the person responsible for the Native American lectures and me. The purpose of the meeting was to integrate and coordinate our respective material as well as to get to know one another. It was decided the meeting would take place in my office.

About a week after this invitation, a solidly built American Indian, whom I guessed to be somewhere in his late forties, knocked on my office door.

After trading introductions, I waved the man called Earth Walks into the brown leather chair across from my desk. Tentatively, not certain exactly what the other was thinking, we went over our respective lecture material, discussing what we would present and how we could fit it together to ensure a smooth flow to the program. In the course of this conversation I made the mistake of mentioning I was going to use one of my favorite quotes, one I commonly used when lecturing on topics such as geological time and plate tectonics.

"And just what quote is this?" Earth Walks asked.

"Society exists by geological consent," I replied. "Subject to change without notice."

"That's a pretty good quote," he said, staring at me with round, black eyes that were so intense I was reminded of an owl staring at a trapped mouse.

"And it's true," I told him. "We all need to pay more attention to geology, to the processes of planet earth. Our society, our very lives depend on it."

"My people pay attention," he earnestly replied. "They always have. What Will Durant said is not completely original. Like many other things, it was borrowed from my people. In 1805 a Shawnee chief called Tekamthi—you might know him as Tecumseh—said:

> The earth is my mother—
> And on her bosom I will repose."

His statement caught me by surprise. Not the Native American quote, but the reference to Will Durant—he knew Will Durant. I thought that unusual.

Looking into his long, almost classical Indian face, I seriously wondered if I should bother pursuing this particular conversation.

Deciding I needed to say something, I sat back in my chair and recalled a piece I had written several months ago. "Those quotes—" I said, "they represent two different times and two very different cultures. But borrowed or not, both statements express the same basic idea: we are all pinned to this blue sphere called earth like butterflies pinned to mounting boards. Our modern society, with its wondrous technology like cellular phones, disposable diapers, computers, and take-out pizzas, is as dependent on the grace and temper of planet earth as your North American Indians were a century or two ago."

Momentarily frowning, Earth Walks turned and looked out the window. Seeming to make a decision, he took a deep breath and turned back towards me.

"Possibly as dependent," he softly said. "Yet there is one important difference. Many North American Indians of that time believed they had a special relationship with planet earth. To them, the earth was family, and they were thus charged with a sacred duty: they were the keepers of the earth. It was this 'special' relationship with planet earth that provided the basis for many of their beliefs and much of their culture and science. It was this relationship that also defined their ideas and practices on how to use and manage their natural resources. These people believed their voices speaking for the land were the voice of the land speaking for itself."

His answer impressed me. Looking at him, dressed in faded jeans and a gray sweatshirt, the old line "you can't judge a book by its cover" echoed in my mind. "Any particular land you are referring to?" I asked.

"Ah," he sighed, giving me a weak smile. "The same one, pretty much, that you will be lecturing about. An area once called the Pays d'en Haut. The French called it this, and it loosely translates to upper country, meaning a land upriver from Montreal. However, for about two hundred years, it also referred to a time when Indian and European were pretty much on equal terms. Geographically the Pays d'en Haut included the Great Lakes region, the Ohio Valley, the prairies, and the Canadian Shield. For the French and Indians this area was a true melting pot, a place where whites would dress in Indian furs, Indians would wear French or English cloth, fur traders might speak several Indian languages, Indians might speak French; and it was a place where intermarriage and cultural respect were not out of the ordinary."

"Until the English came along," I remarked. I couldn't see lords, earls, and ladies being too happy with that kind of arrangement.

"England wasn't really the problem," he answered. "During the time of the fur trade, Indians living in the Pays d'en Haut were actually incorporated into a worldwide industrial system with both the French and the English. In a sense Native Americans were engaged in a wage economy, the basis of which was fashion and protection from the cold. Did you know that over 60 percent of fur trade inventory was actually in cloth, including the famous striped woolen blankets of the Hudson Bay Company, along with floral calicos? And the fur trapped by Native Americans, from beaver, mink, and lynx, found its way to markets in New York, London, Canton, and Paris. But what most people don't know is that even with the great demand for furs, Indians only spent about a quarter of the year working for the fur traders. As Hudson Bay Company governor George Simpson said, they hunt and fish and live as they please the rest of the year."

"If that was the case," I asked, "when did this golden era of Indian-white relationships come to an end?"

"Middle of the nineteenth century," he answered, a touch of bitterness in his voice. "America was booming, and neither the fur trade economy nor the Indian was going to play a part. European-style agriculture, mining, and lumber production drowned out not only the voices of these speakers for the land, but the speakers themselves. For along with these new industries came wagon trains and the so-called iron horse, each bringing thousands of immigrant farmers to free land. And the only reason the land was free is because nobody ever paid us for it. They also never paid us for the great white pine forest they destroyed so they could raise boom towns in the midst of prairie grass."

This said, Earth Walks fell silent. Deciding I didn't really want to pursue this topic any further, I was about to return to the lecture material when he said, "Almost overnight . . . " then stopped, drew in a long breath, then began again. "Indians found their status with America—this new land—drastically changed. During the fur trade they were valued economic partners, providers of food, technology, and geographic and geologic knowledge. Now they were a resented minority that had no place in the new order, or the changed land. Their way of speaking, way of life was headed for the dustbin of history."

"Certainly true and definitely not very nice," I replied. "However, just look at what these new industries have led to, think of what has taken place in this country since that time. Today America stands at the dawn of a new century. And look what a success it has been. Over the past one hundred and fifty years we have indeed taken from the forests, the soil, the earth, and the waters; and with what we found we have built a truly wondrous civilization. The Romans, Mayans, Egyptians, and all the rest were but the bricks and mortar for our glass towers. We have great cities, subways, jet planes, water at the turn of a dial, space probes, and more cars than the Dakotas had buffalo. Minnesota timber built Chicago and St. Louis; ore from the Iron Range won two world wars; and the plains and prairies have helped feed the world."

"You are a geologist," he said, giving me his hard, owl-like look. "A dealer in natural resources. If we are to get along, do this lecture series together, there are some earth truths you need be aware of. First, I suspect you have never opened your eyes to the fact that all this so-called success of America has a price. We, all of us, have trod hard upon the land, or, as Indian philosophers might say, we have shown much disrespect for the only home we have, for Mother Earth.

"Many Indians, including the Ojibwe and Dakota peoples, conceive of time, history, and relationships with the land as a circle, and the circle has come to symbolize a continent-wide American Indian philosophy. Every thought and action in Indian life has repercussions in the natural world, and every event in the natural world, every unusual phenomenon, has deep significance to human lives. We show respect for Mother Earth and she will show respect for us."

I was somewhat taken aback, not only by what he said, but by the fierceness of his voice. "Irrespective of my hindsight," I replied, slowly and a bit sourly, "you might have a point. Your circle of life reminds me of a man named James Lovelock. Ever heard of him?"

"No." Earth Walks said, oblivious to the fact I was just a bit annoyed.

"Lovelock," I said as calmly as I could, "is a scientist, a geologist, and in 1979 he revived and greatly expanded what has come to be called the Gaia hypothesis, named for the ancient Greek goddess of the earth. The Gaia hypothesis is a worldview that has, at its core, the belief that our planet is a dynamic, living place, that there is a subtle connection between all things. Each and every one of us is held and gently tugged by your Mother Earth."

This statement aroused his interest. Looking thoughtful, he turned over a page of his lecture notes.

"Lovelock?" he asked.

"Yes," I said, spelling the name.

A smile crossed his face. "The circle of life to Gaia," he commented. "If what you say is true, then this would be an ancient Indian belief dressed in the white lab coat of a scientific idea."

I laughed at that, and the tension between us dissolved. "Maybe, maybe not."

He nodded and relaxed, settling back into the chair. "Now, the second thing to consider is that together or separate, the circle of life or this Gaia, what these philosophies try to make clear is a simple earth truth. It is a truth you have already hinted at, that what we today call modern civilization is as dependent on our relationship with planet earth as it has ever been. It is the geological and biological processes of planet earth that determine where, how, or even if we continue to exist."

"And it's Native Americans who have always understood this," he continued. "That's why the Irokwa of the northeast believe that one must always consider the results of their deeds on the seventh generation after their own."

Before I could organize any sort of a response, Earth Walks continued. "Perhaps the time has come for us," and here he pointed at me, "you and me, science and the wisdom and observations of Native Americans, to step back several generations and take a new look at some old ideas on how to live with, think about, and share the land we so much depend on and need."

"Did you ever consider," he asked, giving me an appraising look, "ever imagine taking a journey through time? A journey across more than 10,000 years of the Pays d'en Haut to explore its geology, and then try to see how this geology affected the traditions and beliefs, the culture, of the Native Americans who called this land home?"

"Interesting concept," I said, actually thinking that the ancient people he was talking about had nothing in common with modern geology. "However, it might be said that you are just one of those people who can't cope with the modern world, one who waxes nostalgic for the so-called good old days. To go back to the way things were before there was an America."

"By 'those people' you mean Indians?"

"Yes," I replied. "That's what Indians want, isn't it?"

"Absolutely not!" he snapped. "No Indian, unless they're insane, wants that. Let me tell you what Indians want. They want to share the general economic goals of American society. They want to have their historic agreements between their nations and America respected, to have their knowledge and history treated with dignity and respect. What I also want is to help make us all better speakers, not only for this modern land of yours, but for all of planet earth."

"Nice thought," I said, realizing I had hit a sore spot. "And if fate ever decided I take such a journey, where would I begin?"

"You already have," he said. "For such a journey is part geology and part the traditions and culture of Native Americans. For me, such a trip is the blending of the two in the form of stories and oral history for the purposes of demonstrating how early people lived with the land and with Mother Earth. As well, I try and show that old ways of seeing, old ideas and old knowledge are still relevant in today's society.

"For someone like yourself such a journey may be taking the science of geology and seeing what prehistoric Native Americans knew about planet earth. How they used such knowledge, day in and day out, to adapt to and live well with the land. And here you have to remember they lived pretty well with Mother Earth for more than 10,000 years. So they must have been doing something right."

Except for the lecture material and program logistics, that was the essence of our first conversation. Over the next two weeks we spoke a few times on the phone, but it wasn't until our first presentation that I actually started to believe there might be common ground between geology and the culture of Native Americans. Not only that, but I found myself thinking that before Earth Walks, no one, at least to my knowledge, had ever tried bringing the two of them together.

It is for this reason I have attempted to describe some of the events of the past year. In this regard, the first part of the following narrative describes five of the semimonthly lectures and programs Earth Walks and I presented together; the second part describes a trip I took with Earth Walks to the pipestone quarries and a camping trip with him to the north woods to learn about pictographs, petroglyphs, and Native American sky watchers.

So, with imagination and an open mind, I invite you to join Earth Walks and me on a geological and cultural journey across the Pays d'en Haut, a journey that begins where the sidewalk of our society ends and the geology of Mother Earth, along with the culture and heritage of Native Americans, awaits.

CHAPTER TWO

END OF AN ICE AGE

PART ONE

OLD MAN AND THE SPIRIT OF SUMMER

The first lecture Earth Walks and I did together covered the ice age glacial history of the Lake Superior region. Therefore it was mostly my material or, as Earth Walks said, my "show," for I had told him I used lots of slides and overhead transparencies. Earth Walks, however, would introduce the lecture and contribute material where relevant.

During Earth Walks' part of this first presentation, I started to realize that for Native Americans' "living with the land" usually means their way of trying to understand and live with geology. It was this realization that ended up sending me on that long 10,000-year journey with Earth Walks.

The first lecture took place on a Friday evening in the early spring, and the lecture hall was nearly full.

"A lot of potential rock-knockers out there," Earth Walks commented, fully convinced most of the people in the audience were there for the geology.

I disagreed. "It's the Native American aspect of the program they are interested in. The spiritual stuff, the touchy-feely side, that's what they've come for," I told him. This statement received a growl followed by a long, hard look.

"If that's so," he barked, "they have come to the wrong place."

After we had been introduced, Earth Walks, dressed in a beaded Western-style shirt and blue jeans, stepped forward. Spreading his arms wide, he stood motionless, staring out over the audience. Suddenly he brought his hands together in a thunderous clap.

"In the beginning there was ice and wind, and the Pays d'en Haut was silent, for there were no people to hear," he boomed, his voice rising and falling like the swells on a restless sea. "This cold world, however, was set to undergo great changes, changes that would first bring tundra and forest plants, then animals and people; changes that would echo down through the spoken words and stories of Native Americans; stories that began and ended somewhere along the circle:

> *They say this land was covered with ice,*
> *They say nothing lived here,*
> *I say we did,*
> *For our spirits have always been here,*
> *Like the rocks and the waters*
> *We are part of this Earth,*
> *We are part of geology."*

And Earth Walks turned and nodded in my direction.

"If you take a very cold place," I began, as I stepped to the lectern, "add a climate where it snows longer than the baseball season lasts, and stir in a whole lot of time, you have the makings of a glacier. Some twenty thousand years ago, a vast portion of the land went cold and snowy, and with time the continent created a monster of a glacier, a glacier geologists call the Laurentide Ice Sheet."

"In Ojibwe stories," Earth Walks said, coming forward to stand beside me, "the spirit of winter is an old man with long white hair. Powerful and ruthless, he rules the snow-covered and frozen

Old Man Winter and the
Spring of Youth battle for
control of the land.

land. 'I blow my breath,' he boasts, 'and the streams stand still. The water becomes stiff and hard as stone.'

"Then, on a warm wind, a young man arrives at the old man's lodge, walking with light step, in the bloom of youth. Like two shamans engaged in a duel of magic, each tries to overwhelm the other with his powers. After a long battle the young man overcomes the ice and cold of the old man's spells. In victory he cries, 'I breathe and flowers spring up over the plains. My voice calls the birds, and the plants lift their heads out of the warming earth.'

"This story is an allegory of the annual change of the seasons, but it could also reflect a more ancient tradition: an ancestral memory of the continental ice sheet and the ultimate warming of the land."

"To geologists," I said, picking up where Earth Walks had left off, "an ice sheet is not a lake the old man has breathed on; rather it is a mass of ice that happens to be more than a mile thick and a continent wide. A great moving beast that weighs so much it causes the earth's crust to sag downward hundreds to thousands of feet.

"Today our planet has only two continental ice sheets, the Greenland and the Antarctic. The Antarctic glacier is so large it could cover much of Canada and the upper one-third of the United States to an icy depth of more than one mile; twenty thousand years ago, that's exactly what the Laurentide Ice Sheet did.

"A glacier, one of continental proportions, is really nothing more than a great mass of ice on land that has the ability to move under its own weight, much like pancake batter spreads out when poured onto an iron griddle.

"To make ice masses the size of a continent, the earth needs a place cold enough for snow to remain on the ground 365 days of the year, a place where the accumulating snow does nothing but get deeper and deeper. As the fluffy snow accumulates, the weight of all the flakes at the top crunches the zillion upon zillion of flakes at the bottom so tightly together they have all the air pushed out of them. When this happens they turn into something called firn, or corn snow, a dense, granular snow that is much like the snow found along roadsides and sidewalks at winter's end. Firn, with a bit more compression and time, is reborn as clear, crystalline ice.

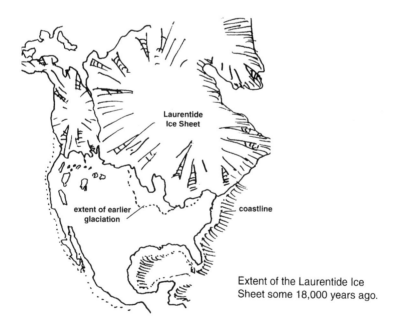

Extent of the Laurentide Ice Sheet some 18,000 years ago.

"In the North Country, some twenty thousand years ago, with snow falling year-round to form rink upon rink of ice, it was only a matter of time before the point of no return was reached. One fine, blizzardy day the growing ice, like Frankenstein's monster, felt big and fit enough to get up and walk the land. The ice, under its own weight, began to ooze, or flow outward, in all directions.

"Normally ice, either as cubes or as sheets, is as brittle as fine crystal. However, when ice has enough weight on top of it—some 165 to 200 feet of snow—it loses its brittleness and begins to take on the look, if not the feel and taste, of pancake batter.

"Glaciers that move like pancake batter are said to flow plastically, and a glacier can flow plastically for hundreds and thousands of miles. In this manner great ice sheets are born, and Earth Walks' old man travels south.

"What geologists call the Pleistocene Ice Age began about 2 million years ago. However, the part of this ice age we are interested in began a mere 20,000 years ago and represents a period of earth history called the Late Wisconsin. For the last time the massive lump of ice centered over Hudson Bay plastically flowed outward to cover all of Canada and the upper one-third of the United States. This frozen wilderness extended from the Atlantic Seaboard across the southern Great Lakes into the prairies of Alberta. The ice sheet reached its maximum size about 14,000 years ago, when it covered an area of about 45 million square miles. Tongues of ice extended as far south as central Illinois, Missouri, and Kansas and into southern Ohio and Indiana.

"As the ice flowed over the land, whatever vegetation had been present either wilted away or was simply freeze-dried. With the coming of the cold and snow, tundra conditions predominated, and this included permanently frozen ground (permafrost) and intense frost heaving. However, just as the tundra mosses and grasses were getting nicely settled in, along came the legendary old man and his icy breath, and the tundra was forced to shift itself south. This happened over and over, again and again, as the ice expanded and the vegetation retreated.

"Glaciers are not benign beings that simply cover the land and then wait to melt. Instead, they are some of the hardest workers Mother Earth employs. They work 365 days a year to bulldoze high hills until they are as flat as the state of Iowa, to dig valleys deep and long, and to change the course of rivers and create new ones as easily as a wave washes away a child's sand castle. They bury the land under a thick blanket of 'earthy' debris and constantly grind, crush, and smash rocks.

"But ice alone, either as part of a thick glacier or as a thin cover of crystals on a pond, is way too soft to grind, smash, or dig anything. In fact plastically flowing ice has the consistency of caramel syrup. If you take caramel syrup and pour it over large slices of red Delicious apples, the syrup slowly oozes between the pieces, filling up all the nooks, cracks, and crannies. If enough syrup is poured onto the apples, it eventually covers them like snow blankets flowers. But the syrup doesn't smash, dig out, or grind up a single piece.

"So if the ice doesn't do it, who is the guilty party? It turns out it's the hitchhikers the ice picks up on its journey. Glaciers have a bad habit of picking up any loose rocks that are found on the ground they travel over, as well as taking rocks off the sides of valleys and basins. These rocks become frozen solidly into the ice, turning the glacier into something like rough sandpaper. It is these rocks that give the glacier its scouring, digging, and tunneling power, power that geologists call abrasion. Fine parallel scratches and grooves that remain on bedrock in glaciated areas are called striations. They are the common graffiti left by these rocky hitchhikers."

"And you know," Earth Walks said, coming forward to take the slide-changer from me, "Early American artists recognized a helping hand when they saw one. They used glacial striations as elements in the making of petroglyphs, probably because the rocks were so hard to work that any premade mark was better than none. So a groove became an arm, a striation one side of a body or an antler on an elk. Paint these up, and it's hard to tell the glacier's work from that of the human artist."

"The continuous grinding and scouring action of a glacier," I smoothly continued, moving to the next slide, "slowly breaks the rocks it carries into smaller and smaller pieces until they are reduced to powder-size material called rock flour. Most glaciers produce so much rock flour that meltwater leaving the ice has a milky white appearance, similar to skim milk. Some North American Indians call this water 'old man's blood.'

"Glaciers also pick up rocks by a process called plucking. Plucking occurs where small amounts of meltwater seep into broken and fractured rocks at the base of the ice and freeze solid. This frozen water firmly attaches the rock, be it baseball- or garage-size, to the bottom of the glacier. As the ice flows forward, the stuck rock is pulled out of the ground as easily as a loose tooth by a peanut butter sandwich. Over a long period of time this plucking action creates rock basins and bowls, which, when the ice eventually melts, become filled with water to form everything from ponds to Great Lakes.

"Continent-size glaciers don't simply rampage across the land, destroying everything that happens to get in their way. Instead, they carefully select how and where they want to go, and more often than not they have favorite landforms they like to follow, and rocks they enjoy destroying.

"In this regard glaciers are a lot like wolves. A pack of wolves will often single out and attack the weak, the old, and the sick; glaciers attack, scoop out, and carry away the soft, the layered, and the broken. Just the type of rocks that are easy to erode and to move.

"Glaciers also take advantage of topographic differences they come across on their travels. They simply love low areas; be it a valley, basin, or some other depression, they will follow it every chance they get. Which is exactly what the Laurentide Ice Sheet did when it came across a long, topographic trough in Minnesota called the Superior-Minnesota Lowland. For the Laurentide Ice Sheet, this partly eroded depression was like finding a six-lane freeway in the middle of the mountains; the ice flowed into and down it all the way to central Iowa.

"For more than 7,000 years, in these varied ways, the Laurentide Ice Sheet moved across, eroded, and covered the land."

I paused for a moment to organize my thoughts, then began the second part of the glacial story.

"Like the young spirit in the Ojibwe story," I said, "one fine spring day warmth and sun came to the land and stayed. Approximately 13,000 years ago the temperature rose, and the planet started to warm up. With rising temperatures more ice melted than formed, snowfalls decreased dramatically, and April could again look forward to having showers for May's flowers. Like a badly wounded animal, the Laurentide Ice Sheet slowly melted back to its polar den.

"With time the ice disappeared, but it left behind a lot of the rock baggage it had carried with it. The ice sheet left a vast landfill of sediment and rocks, tens to hundreds of feet thick. This material leveled out the countryside to make mile after rolling mile of rocky soil, a landscape dotted with thousands of small and large lakes. To go along with the lakes were brand new valleys, new rivers to flow down them, and new hills to climb alongside them.

"A glacier gets rid of its possessions in one of two ways. First, it simply gets up and goes, leaving a mess behind. The glacier melts away, and a house-size boulder ends up sitting on a cliff edge, or in the middle of a swamp.

"The other way a glacier disposes of its stuff is by the sweat of its own blood. Water from the melting ice carries ton after ton of sediment and rock away from the dying beast, and deposits this across the land.

"These two disposal methods create a huge variety of landforms and landfills that, for thousands of years, the North American Indians lived with, used, and tried to give meaning to. We'll look now at some of the more common ones.

"Glacial 'erratics' are large boulders, some bigger than a house, carried far from where planet earth made them. These are the 'how'd that get there?' rocks, and they are magical and spiritually significant rocks to the North American Indian."

Pausing, I handed the slide-changer to Earth Walks and stepped aside.

"Native Americans knew enough geology to realize these lonesome boulders did not belong to the land they were found in, that they were different from the rocks that made up the surrounding landscape," he explained. "To them these were special rocks, gifts of the gods, powerful spirits to honor and sometimes fear.

"For example, the Pipestone quarries in Rock County, Minnesota, are places long held as sacred ground by North American Indians." And here he showed a slide that illustrated the location of the quarries. "The bedrock is a hard, resistant pink material called Sioux Quartzite. Sitting just outside the sacred quarries are three large, grey boulders of an igneous rock called granite. These granite boulders are strikingly different from the pink quartzite, and are obvious visitors to the area. Therefore, they must be special rocks, the rocks of myth.

"One myth explains how long ago, all of the people were destroyed in a terrible war. All except three Indian maidens who fled to these rocks and were saved by the Great Spirit. So all of today's Indians are descendants of the three maidens, who live on in the form of the three boulders.

The three Indian maidens of the Pipestone quarries.

"Or possibly it was the war eagle who laid three large eggs to mark the spot of the quarry. Three maidens were chosen to guard the eggs and the sacred quarry. The maidens lived beneath the rocks so they could watch all who came near. Indians left gifts near the rocks as offerings to the maidens and to make sure they would have good quarrying. Some Indians left only their tracks, as they were too frightened to get very close and hurled their gifts of tobacco or food at the boulders. Braver Indians, however, ventured close enough to carve turtle petroglyphs into the gray rocks.

"Then there is a village in Dakota County, Minnesota, called Red Rock. It gets its name from a large boulder of granite that, long before the French and English came to this country, was kept freshly painted with bright red ochre pigment. This was done to honor the presence of this special visitor to a land where it clearly did not belong.

"It turns out Europeans, as well as Indians, noted these special rocks. There is a twenty-ton boulder of dark green basalt that sits out on the flood plain of the Chippewa River in southwest Minnesota. Years ago local folk recognized that it was completely different from the pink granite that makes up the local bedrock. Knowing that this was a stranger from some strange land, the locals came to believe it was a visitor from outer space, and so named it the Montevideo Meteorite. The Native Americans also knew about this strange erratic. They had honored it for hundreds of years by rubbing it with buffalo skins to keep it clean and shiny."

"Next, for geologists," I continued, showing the next slide, "comes 'till,' a wonderful Scottish word for stiff, rocky ground. Till refers to sediment and rocks deposited directly by melting ice. Deposits of till are not layered (bedded), and they are said to be poorly sorted because rocks of all different sizes occur next door to each other.

"Landforms constructed of till are called moraines, and these come in two main flavors—rocky road and heavenly hash. Rocky

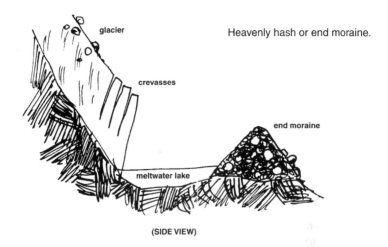

(SIDE VIEW)

road moraines, also called ground moraines, are broad blankets of ice-deposited till. Tens to hundreds of feet thick, these fill in depressions and smooth out the landscape. When the North American Indians or Scottish farmers dug into them, they quickly found out why they are called rocky roads.

"Heavenly hash moraines, also called end moraines, are ridges of till that form at the edge of a melting glacier. These form only if the glacier has been stuck in the same place for a long period of time. At such stuck or stagnant ice fronts, the glacier continues to bring in lots of hash—a jumbled mixture of rocky debris and finer sediment—which, as the ice melts away, is left high and dry in the form of long, irregular ridges (parallel to the ice front). These can be tens of feet high and miles long and are better dams than the giant Pleistocene beaver ever made. A special type of stuck moraine is a terminal moraine, which is just that: the end of the line for a glacier. These are ridges of till that mark the spot of a glacier's furthest advance.

"If the ice manages to surge forward again, it might run over one of its own heavenly hash moraines. In doing so, the ice reshapes the moraine into an elongate, streamlined hill called a drumlin.

"Melting glaciers also let loose enormous volumes of water that carry away and deposit sedimentary material. Water-deposited glacial sediments are called outwash. Unlike till,

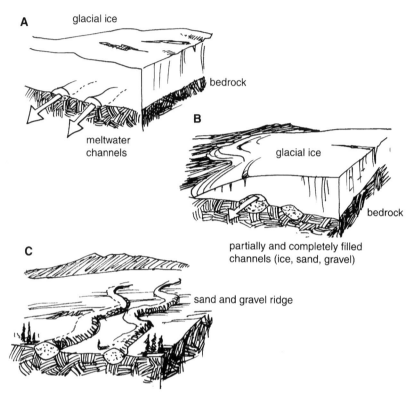

Formation of eskers: (A) meltwater channels form at the base of a melting glacier, (B) the channels slowly become filled with sand and gravel, (C) ice melts away, leaving behind "snakelike" ridges, which represent the filled meltwater channels.

outwash is layered (bedded) much like a lasagna, and the pieces of rock found within each layer are said to be sorted, because they are all about the same size.

"Landforms created by outwash include eskers, kettles, kames, and loess. Eskers are snakelike ridges of sand and gravel formed by meltwater streams flowing in tunnels within, and at the base, of a melting glacier.

"Kettles are small lakes that form at the margins of fast-melting glaciers. When glaciers undergo rapid meltdown, they tend to leave behind large blocks of ice. These blocks become stuck and isolated in the glacial till, and when they finally melt away, they leave behind a hole that can trap water to form a lake or pond. Kettle lakes can be miles wide but are typically no more than fifty feet deep. Minnesota could easily have been called the land of 10,000 kettles, and many a self-respecting prairie pothole is really a kettle lake in disguise.

"Kames are steep-sided, conical hills composed of sand and gravel. They form when sediment, deposited from meltwater, fills holes in the stuck or stagnant ice at the glacial front. The making of kames is much like the making of muffins. Muffins are often baked in muffin tins, which have six to eight round holes in them to hold the muffins. Think of the holes in the muffin tin as the holes in the glacial ice. The batter, or sediment, poured into the holes fills them up, and when they are baked and taken out of the pan—the ice melts away—you are left with a muffin shaped hill."

"And you know," Earth Walks added, suddenly at my side, "Kames are often next-door neighbors to kettles, and both were important landforms for the North American Indians. Kettles and prairie potholes were natural magnets for all sorts of animals, especially in the spring and during dry spells. For the Indians, kames were ideal places to observe the movements of these animals, and provided hiding places for a hunter to await the proper moment for a strike."

"Last comes loess," I continued, as Earth Walks left the stage, "the fine-grained sediment picked up by the wind from glacial outwash surfaces and deposited elsewhere. Loess deposits, if thick enough, as in the western and southern parts of Minnesota, form rich topsoil. A modern-day farmer's paradise, it was also heaven on earth for the North American Indians some 1,000 years ago. They planted and harvested a type of grain called maize, which thrived in the fertile, well-drained loess of Minnesota.

"When the end of the age of ice finally came, it was fast and furious, geologically speaking. On its last trip south, the Laurentide Ice Sheet sent lobes of ice on a visit to central Iowa and southern Minnesota. This took place about 14,000 years ago, and a mere 2,000 years later, both of these ice lobes had melted back beyond the Canadian border.

"The rapid melting of all this ice caused sea levels to rise an average of four hundred feet worldwide. Meanwhile, miles from the ocean, in the heartland of America, all basins, ponds, and gopher holes were filled to the brim with glacial meltwater. Thousands of lakes came into existence, lakes that varied from an acre or two to more than fifty thousand square miles. Many of these larger glacial lakes lasted no more than a few thousand years. However, their size and long, sandy beaches greatly affected the people who lived beside them, as well as the postglacial landscape, for ages to come.

"When you fill a depression with water to create a lake, you almost always create an outlet to go along with it. These lake outlets range from mere trickles that dry up in the summer to mighty rivers that flow to the sea—rivers such as the Mississippi, Ohio, Missouri, Red, and St. Croix, all created because of an ice sheet called Laurentide.

"The largest of all the glacial lakes came to be called Agassiz, named after the Swiss naturalist Louis Agassiz, who in 1837 first proposed the idea of a Great Ice Age. Lake Agassiz was born about 12,000 years ago, when the ice melted off Iowa and most of Minnesota, retreating behind a topographically high area that today forms the southern part of the Laurentian Divide. The

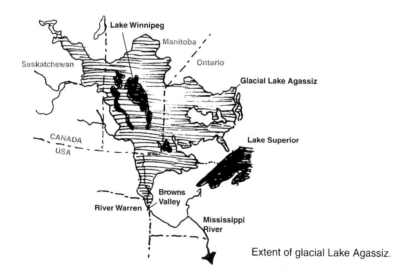

Extent of glacial Lake Agassiz.

Laurentian Divide eventually became a continental divide separating north-flowing waters, such as those of the Red River, from south-flowing waters, like those of the Minnesota and Mississippi Rivers.

"Lake Agassiz formed, then grew, as meltwater became trapped between this natural dam and the north-retreating edge of the melting ice. Various levels of the lake are marked by a series of north-south beaches, some of which can be traced for hundreds of miles. It is within these beach deposits that artifacts and bones of the area's oldest human population are found, which just goes to prove that even back then beaches were great places to be. The distribution of old shoreline features and lake-bed sediments indicates that Lake Agassiz covered more than 125,000 square miles, though at any one time it is believed the lake was no more than 50,000 square miles in size. As small as this may sound, it was still considerably larger than present-day Lake Superior.

"The climate continued to warm up, and the ice melted like ice cream in the sun. Large quantities of melting ice produced huge amounts of running water, and this water ran mostly into Lake Agassiz. This caused lake levels to rise faster than orange prices after a frost in Florida. Agassiz could only hold so much water, and therefore it was only a matter of time before something had to go. What went was the Laurentian Divide near what is now Browns Valley, Minnesota. The rising water broke through and poured over the natural dam and, as free, flowing floodwaters will do, proceeded to dismantle the high ridge, sand grain by boulder by grass blade. In less time than it takes to run the Boston marathon, the water had excavated a hole large enough to hold five 'chunnels,' and Lake Agassiz poured through like a high-speed train heading for Paris.

"This great deluge surged over the land. By following topographically low areas, it etched out a course that would soon—geologically speaking—become a trench some 250 feet deep, 250 miles long, and, in places, more than 5 miles wide. Year after year, between 9,500 and 11,500 years ago, great floods poured out of Agassiz to rock, rattle, and drench the land. The deep valley formed by these waters is known today as the Minnesota River Valley, and the waters that raged along it were part of a river called Warren.

"River Warren, before it was finished, cut down through 3.6 billion years of earth history to uncover the oldest rocks in Minnesota, and some of the oldest rocks in the world. Not only did River Warren expose rocks called the Morton Gneiss to the light of day, but it cut large holes into them—called potholes—with whirlpools of water that were strong enough to turn boulders into rock drills. The water spun the boulders around and around, faster and faster, year after year, down and through the Morton Gneiss. The flood waters also battered and eroded the rocks, leaving lonely, isolated islands and steep, ragged knobs strewn across the valley floor.

"Features present in the Minnesota River Valley, and in the Morton Gneiss, are similar to features observed in other areas where catastrophic flooding is indicated. For example, thirteen thousand years ago the ice dam that blocked the Clark Fork River Valley in Montana collapsed, and glacial Lake Missoula was drained away in one great surge. Lake Missoula was approximately the size of Lake Michigan and is estimated to have had a water depth of about one thousand feet. If these numbers are anywhere close to the real thing, then the hole in the ice dam sent more water into eastern Washington, from Spokane to Portland, than the combined flow of all rivers in the world today. The remains of this monster flood can today be seen throughout what is called the channeled Scabland area of the United States.

"Flood days at Agassiz, and its daughter, River Warren, ebbed and flowed with the turning of the seasons. Water volumes were high and fierce in the spring and early summer, and low and lazy in the fall.

"Whether raging, down-cutting, and destroying or lazily gurgling along, River Warren emptied into the Mississippi River near what is today called Fort Snelling in Minneapolis. During flood stages the water of Warren entered the Mississippi with so much power, it took huge divots out of the bottom of the Father of Rivers. This left the Mississippi north of Fort Snelling some thirty to sixty feet higher then the waters to the south. With nowhere to go but down, the northern part of the river created a waterfall that, over the past 11,500 years, has eroded back to a place called St. Anthony Falls.

"Glacial River Warren happily destroyed the land for 2,000 years, until the retreating ice uncovered a topographically lower drain, which allowed Lake Agassiz to flow northeast, into Lake Nipigon and then on into Lake Superior. Almost overnight the mighty River Warren became the wimpy Minnesota River, and a violent chapter in the geological history of Minnesota came to an abrupt end.

"Lake Agassiz became a shadow of its former self, and by 8000 B.P. (Before Present) was gone. Gone, but not forgotten, for there remain old beaches, lake sediments, shorelines, and a few smaller lakes to remind us of what was, and what could be again. Present-day lakes like Winnipeg, Lac La Ronge, Athabasca, Red Lake, and Lake of the Woods all occupy parts of Agassiz's old basin. These lakes provided the Indians and voyageurs with an interconnected water route into the North American interior. The lakes and rivers that the Laurentide Ice Sheet left behind allowed the Indians to establish large trade networks that crossed the continent and made the fur trade a successful venture."

This ended the first part of the program. We had a fifteen-minute coffee break and general question period, at which Earth Walks was conspicuous by his absence. I was soon to find out why.

PART TWO

WHEN THE GREAT PANTHER RISES

When the audience was again seated, Earth Walks came onto the lecture platform wearing a long, fur-lined deerskin jacket tied with a fur sash. Beneath this he wore leather leggings; on his feet were a pair of deerskin boots tied with leather thongs. He had taken the braids out of his long, black hair so it hung loose, reaching to his shoulders, and he carried a wooden spear.

Native Americans were there to witness River Warren's greatness and power.

"What you may not know," he said, staring over the audience, "is that Native Americans were there to see Lake Agassiz's greatness and witness River Warren's power. Imagine what it must have been like to stand on the edge of the Minnesota River Valley some 10,000 years ago. What a sight River Warren must have been. The catastrophic flood of water, crashing and raging like five hundred Niagaras. Think of the stories, myths, and legends that must have been born and told, only to be retold around campfires down through the ages.

"And how about the people who lived there when it all came to a tragic end some 9,500 years ago? Together, let us try and imagine what that spring may have been like when the river refused to flood, when the lake of lakes virtually vanished from the land. To do this we will travel back in time and visit with a man called Peetwaniquot, one of a group of people who spent summers on Agassiz's sandy shores and winters in Warren's sheltered valley. Through him I will try to show you what that geological and cultural catastrophe may have been like."

Earth Walks then walked over to the chair he had placed in the center of the stage and sat down.

Imagine it's spring, approximately 9,500 years ago, somewhere along the River Warren. When we first see Peetwaniquot, he is sitting on a large outcrop of Morton Gneiss, staring into a deep, wide valley.

The muddy, slow-moving river he watches has him worried, more so than he would ever say. What keeps running through his mind are the words of his father, who had heard them from his father, and so it had been for all of time.

> *When the Great Panther rises in the evening,*
> *And the Fisher stars swing overhead, releasing the birds,*
> *When Father Sun looks into our lodge doorways in the*
>   *morning,*
> *Then we must strike our camp,*
> *Go to the ridge tops*
> *And wait until the tall waters pass,*
> *Until the land is swept clean and born anew,*
> *Then will be the time to receive the gifts of the great water.*
> *Until now.*

The Great Panther had risen to fly across the sky. With his coming the afternoon sun sent the snow running to the nearest streams, and from there on to the rivers. As they had done forever, the people packed the winter camp and moved to the top of the valley. To the highest place where the forest ended and the grassy slope began. There they set the lodges, made the fires, and began the wait.

And wait they had, longer than ever before.

Something was terribly wrong. Warmth was all over the land, small pink and blue flowers poked up through the grass, the forest was alive with birds, and all that remained of the Old Man were hard, crusty patches of snow spread across the darkest part of the forest floor. And the river, what a sight it was. Never before had anyone seen it so small, a slow, muddy gurgle you could wade across if you were tall enough. It was unbelievable. Where was the mad rush of water, the banging and clatter of the lodge-size boulders, the deep, foaming pools full of spinning trees and the broken bodies of animals caught in the open of the valley floor?

Peetwaniquot had spoken out, saying there were great changes, he could smell them in the air and taste them in the water. A few of the others said the people had done an evil thing and were once again being punished by Gichi Manidoo. But no one could think of what they had done, or why they would deserve such a terrible fate. Most of the people said wait; the water would come as an elk came to the patient hunter. What they all agreed on was that the answer might lie with the great lake; who would be the one to go north to see?

They had waited for him to answer. To stand and take the pipe and say he, Peetwaniquot, the boatman, would go. It was his place. Better than anyone he knew the pulse of the lake, the islands and beaches, the fishing spots and where the dangerous rocks lay. He was the first to know when a storm was building and when the waters would change from cold to warm. His grandfather had been the boatman before him, and had taken him on the great journey to the blue ice to prepare him for this honor and duty.

The Great Panther had risen and set since he had taken the clay pipe and given his speech. Now, far from the village, he moved north through the pine woods.

As he traveled through the twisting valley, one ear listening to the river and the other to the woods, he thought about the people and the old stories. He remembered the first time he had heard the words, and then reheard them, over and over until they were burned into his very being.

It was said, his father had told him, that long ago the long noses passed out of the land, and Manidoo sent the water to punish the people.

The people had followed the great herds to this valley, and here they found more game, fish, and plants then had ever been seen before. But what they had come for, what had led them north to this remote place, were the tusked ones, giants with noses so long they could be swung like clubs, and one blow was enough to kill. These great creatures stood two or more people tall, and it was said their roaring caused the entire forest to shake. The meat from a single animal fed all the people for more than half a moon.

Life was good and easy. The long noses could be trapped in bogs, or driven over a cliff. Great ceremonies of thanksgiving would then be held for the great master, the Ogimaa of the long noses. But then something very strange and terrifying happened. A group of the mighty beasts were driven into a shallow bay, and the entire herd followed.

They all perished in the mud and water. Butchering was swift and skillful, but the whole people could not keep up with the spoilage and decay. Most of the great herd fed the ravens, vultures, and a few brave wolves. And these were the last of the long noses in the valley!

A feeling of deep foreboding fell upon the people. Surely Manidoo would punish them for this great sin. And it was so. The tall waters came like an avenging spirit, and many died. Those that escaped heard the words of Gichi Manidoo. He told them that when the Great Panther appeared in the night sky, the great waters would come. The waters would remind the people of their greed and of their duty to all creatures who lived on the land. The water was to be a warning of what would happen if they again abused Mother Earth.

Since that time the people had watched for the Great Panther. With its rising they left the valley, and from the highest place waited and watched for the coming of the high waters.

Peetwaniquot thought of last year's passing: the brown raging water, rising like the moon higher and higher up the valley sides, and the banging and clattering and roaring as it rushed past. The noise had drowned out their singing, and they had stood and watched in silence. With the passing of the high water, the spirit celebrations had begun. There were promises to the land, the final feast of winter, and much dancing and singing.

He thought about all this, but still it didn't stop the fear from filling his mind as he passed through the Wolf's Bite, the boulder-filled gash sliced out of the tall ridge by the terrible waters. Once through the notch he climbed a low hill of fine, yellow sand and stood looking out, over the great lake.

Gazing north, he dropped his spear and fell to the ground. There he lay, as frozen as the blue ice, and blind, for what he thought he saw was impossible. Manidoo could not be so cruel.

Yet the sandy beach was there, he could see it from where he lay, and behind it sat the sacred rocks painted red and silver. He could still see the places where the summer lodges had stood, and the snakelike ridge of sand and rock where the dead were put so they could begin the long journey to the spirit lake. But the rest—impossible. Where the boats were kept and the fish cleaned, there was nothing but sand and cracked mud. The mud extended to the north as far as he could see. The lake had left the land, the great water spirit had deserted the people, and nothing would ever be the same again. Sitting up, he hugged his knees and began to weep. He wept for the people, and he wept for all those that would come after him. What had they done? Why was Manidoo so angry?

The applause began almost before Earth Walks was finished speaking. Rising to his feet he held up his hand to stop it. "We all have something to learn from Peetwaniquot and his people," he shouted. "First, Peetwaniquot didn't lie on that beach and die. Oh, no. He picked himself up, brushed the sand from his body, wiped away his tears, and went back to tell his people. For them, as for him, it was a calamity; the world as they had known it was over. But did they lie down and die? For the second time—absolutely not! They, as a people, adapted to this radical, life-threatening change; they learned from this and moved on. And one of the important lessons they learned, one they have passed down to us through the ages, is that when Mother Earth takes something away in one geological event, one geological moment, she gives something back in return. Peetwaniquot and his people lost Lake Agassiz and River Warren because the Laurentide Ice Sheet melted back to uncover a lower outlet, an easier drain for the great lake. But in return for this great loss the people were given wide, navigable rivers like the Mississippi and Missouri, and along these rivers Mother Earth had provided wetlands, marshes, small lakes, all of which teemed with fish, waterfowl, and game. And so the people moved to the river environment, they adapted, improved, and prospered. We would all do well to remember this."

Inclining his head slightly toward the audience, Earth Walks, to loud applause, turned and strode from the stage.

CHAPTER THREE

PEOPLE OF THE
PAYS D'EN HAUT

PART ONE

ORIGINS

The second lecture, which was primarily concerned with the earliest inhabitants of the Pays d'en Haut, also took place on a Friday evening. On this night the lecture hall was full, and I told Earth Walks that word of his storytelling prowess had spread far and wide, that all these people had come in hopes of a repeat performance.

"They won't be disappointed," he matter-of-factly said.

At the appointed time I stepped to the lectern, welcomed everyone, and began the introduction.

"What do geologists and archeologists really know about the people who lived on the shores of Lake Agassiz and along the banks of River Warren, or, for that matter, the people that came before and after them?" I asked.

"As it turns out, much of what we know about early Native Americans is in the form of shadow clouds, mists on a vast lake of time whose essence is but the landfills, cemeteries (middens), fire pits, and dirt floors of the past.

"To attempt to give form and substance to these shadow clouds, archeologists and geologists try to reconstruct the cultural and geological environments of Native Americans from the end of the ice age to the coming of the Europeans. Unfortunately this is much like trying to put a jigsaw puzzle together with 70 percent of the pieces missing. The pieces that do exist come mainly in the shape of

(1) bones of fossil animals, many of which are now extinct,
(2) landforms, the processes that form them, and their changes through time,
(3) remains of fossil plants and trees, as well as ancient seeds and pollen,
(4) the stuff of garbage dumps, burial sites, fire pits and dirt floors."

"And last, but possibly most important," cried Earth Walks from behind me, "from the relatives of these early peoples—the North American Indians themselves.

"This last point is important," he exclaimed, taking my place at the lectern. "Traditions, myths, and oral histories of modern Indian societies are important sources and reference points in trying to understand the people of the distant past. Archeologist Ronald Mason has stated that, and here I quote, 'in more ways than technology the Great Lakes Indians were stone age people' just prior to the coming of the French and the English. In other words, archeologists believed that Native Americans living in the sixteen and seventeen hundreds had not changed much from their early ancestors.

"Unfortunately, for many of us, the term 'stone age' conjures up images of rubbing two sticks together to make a fire, or of people who live in caves, carry big clubs, and communicate by means of shoulder shrugs and low-pitched grunts. Early Native Americans, however, were as far removed from this image of the stone age as we are from ox carts, candlelight, and quill pens. The authors of *The Native Americans* point out that the first people to arrive in the Pays d'en Haut were, and again I quote, 'fully developed modern human beings' who"—here Earth Walks picked up a book from the lectern and read:

> *brought with them an ice age patrimony, including many basic human skills: fire making, flint knapping, and effective ways to feed, shelter, and clothe themselves. They lived in close knit kin groups, enjoyed social interactions, and shared beliefs about magic and the supernatural. They spoke a fully human language. They lived in diverse and sometimes unstable environments. Over the generations the first Americans confronted and solved colossal challenges.*

"Colossal challenges served up by Mother Earth," he thundered, actually pounding the lectern. "Hard and fast like a professional tennis player trying to hit an ace. But these challenges were met, they were confronted and solved. How did Native Americans do it? How did they manage not only to survive, but to flourish and grow?

"One important factor is that this so-called stone age was not really a stone age at all. It was, in fact, an age of bone-tooth-copper-antler-wood-bark-hide-plant-clay-and-sinew as well as stone. But, most importantly, it was an age of information. Most people today would be hard-pressed to name twenty wild plants.

Coming of the people to the Pays d'en Haut.

The historic Ojibwe knew more than two hundred. Each adult knew which plants to eat, paint with, or use as medicine; when to pick it, whether to use the root, stem, leaf, flower, fruit, or seed; whether to make a tea with it, a poultice, or a seasoning in a stew. And Native Americans possessed such knowledge for thousands of years.

"Overall," he passionately cried, "their knowledge of the land and its plants and animals was encyclopedic, their myths about the land and themselves would fill a bible, their social organizations were complex, and their sense of community and mutual obligations would serve many of us quite well. These people, when all is said and done, had the basic technical, scientific, and social framework to survive, adapt, and grow.

"Scientists," and here he glanced at me, "try to reach far back in time by looking at broken spear points lodged in bison bones, or by examining shards of pottery found in a midden. However, to begin to comprehend that distant past, it must be seen as more than landfills and kill sites and burial mounds. There is flesh and blood, spiritual awareness, fear and dreams, myths, curiosity, and scientific and technological knowledge. It should be remembered that these people not only adapted to the changes around them, but they did it very well. In the circle of time, between the edges of the ice and the destruction of the buffalo, are ten thousand years of living well with the land—physically and spiritually."

Here Earth Walks paused to take a sip of water and consult a page of notes.

"With the melting of the ice," he began again, his voice low at first, then rising louder and deeper, like a coming earthquake, "bare ground and rocks were exposed to the sun and air for the first time in thousands of years. The land started to dry out, and the frozen ground slowly thawed. With the sun and warmth came the first plants and trees, and these had to be as hardy as the Native Americans themselves. These first pioneers came with plant names like moss, lichen, sedge, and short-root grass, and they formed vast stretches of a land called tundra. The first trees came mainly with the name of spruce, and they formed thick, green forests.

Native Americans followed herds of woolly mammoths north to the Pays d'en Haut.

"With the coming of this vegetation, so came great herds of migrating animals, and none of these were quite as great as the tundra-loving woolly mammoths and the spruce-chomping mastodons. And hot on their trail, just over the horizon and through the woods, were the first humans, the ancestors of Peetwaniquot and his Lake Agassiz people—people called Clovis-Folsom by archeologists. These elephant hunters arrived in the Pays d'en Haut between 11,000 and 12,000 years ago.

"Who were these first people? And where in the world did they come from? A generation ago most scholars believed humans arrived in the Americas by way of the Bering Strait land bridge some 12,000 to 13,000 years ago. This area became high and dry when the sea between Siberia and Alaska was exposed by dropping water levels, due to the formation of glacial ice. These first Americans walked across this land bridge, following herds of caribou, musk-ox, and woolly mammoths; and they kept right on walking until they reached the southern tip of Chile. According to these same scholars, the continent's first marathon walkers multiplied rapidly, and in about 1,000 years they not only covered two continents, but also managed to find the time to form the many different cultures of Indian America!

"Indians, including the modern-day Dakota and Ojibwe, have never accepted this idea. Every Indian culture has their own earth creation myth and their own myths of the first appearance of themselves as a people.

"The Hopi, for example, describe successive worlds being destroyed in turn by fire, ice, and flood due to the sins of their ancestors. Interestingly enough, their western homeland exhibits all of this geology: volcanic eruptions and lava flows, mountain glaciation, and intermountain damming and flooding. Finally the Hopi make an ocean crossing over their flooded world, they stop at tropical islands, and during the last part of their journey they have to paddle uphill to finally arrive at a great wall of western mountains. These they cross to reach their desert sanctuary, a land they were given by their creator—a harsh land that will test their faith, a land now called Arizona.

"The Cheyenne describe an ocean crossing in round, hide-covered boats. The Nootka, consummate seafarers still, also relate an ocean crossing. The Kiowa, on the other hand, emerged from the earth through a great hollow log.

"In the Ojibwe version the Great Spirit created the first earth, and brought forests, flowers, and animals into being. Then came great destruction in the form of a cataclysmic flood, and all but the water creatures were destroyed, and the world remained a sea for many centuries.

"Skywoman lived in the heavens, alone and desolate. The water creatures, taking pity on her, persuaded a great turtle to rise up from the waves and offer his back to her as a sanctuary. Skywoman descended on a flight of great Canadian geese. Settling onto the turtle's back, she asked various water creatures to try and get her some soil from the bottom of the sea. Many tried, and all failed except the humble muskrat. He floated to the surface, exhausted from his great dive, but with the precious earth clutched in his claws. Skywoman breathed on the soil, and it grew, sprouted plants, and spread over turtle's back to form an island. This island was called Mishimakinaong, know today as Michilimakinac.

"Skywoman then gave birth to a boy and a girl, the first people who were called the Anishinabeg, beings made out of nothing, spontaneous beings.

"In most Indian creation myths there appears to be an awareness of geological processes and an emergence into a new land. One can read them as allegories, or as a reflection of the Indians' place in the cosmos and their long tenure on the land. There is no common myth, and none mentions the crossing of a great land bridge, the Bering Strait.

"The great variation in the myths is easily explained; there were five hundred Indian nations in North America alone, with five hundred different languages from at least seventeen linguistic families. These were so different that if an Ojibwe-speaking Algonkian met a Dakota-speaking Siouxian, it would be like an English-speaking person trying to strike up a conversation with someone who spoke Mandarin Chinese.

"It would appear that archeologists and anthropologists have paid little attention to Indian origin myths in creating the Bering Strait theory. This theory of a great land bridge may have been first suggested by the Catholic Church shortly after Columbus' voyage, and it served as an explanation for the existence of a people not accounted for in scripture. According to Sioux author Vine Deloria, the Bering land bridge theory does illustrate a constant theme, and here I quote: 'A good many scientific and/ or scholarly beliefs about Indians originated as religious doctrines. As religion lost its influence as an opinion maker, the idea was picked up by some secular scholars, transformed into scientific theory, and published as orthodox science.'

"There are other problems with the land bridge theory. Chukchi peoples of Siberia can visit their cousins, the Inuits of Alaska, by paddling across the Bering Strait, and, as recent evidence indicates, they could have done so for more than 60,000 years. These people did not need dry land, nor did they have to walk. Some years, when it is covered by pack ice, the Bering

Great turtle offered his back to
Skywoman as a sanctuary.

Strait can be crossed by dogsled, as was demonstrated recently by Paul Schurke and his Inuit companions.

"Though some scholars, over the last twenty-five years, have still held to the crossing of the land bridge some 12,000 to 13,000 years ago and the establishment of the first major new world culture at Clovis New Mexico around 10,900–11,200 B.P., there are now many criticisms and variations offered. Some of these my geological colleague will now tell you about."

"Writer James Adovasio," I said, rising from my seat and coming to the lectern, "has pointed out that humans don't sprint across their environment, which the first Americans would have to have done to create the Monte Verde site in Chile. Tom Dillehay, who excavated the Monte Verde site, found stone tools, bones, mastodon meat, wild potatoes, and twenty types of medicinal plants, and the site dates from 12,300 to 12,800 years ago. This has led some archeologists to describe the trek from the Bering Strait to Monte Verde as the 7,500-mile dash. Dillehay also believes that an older, lower level at the Monte Verde site contains charcoal that may be nearly 33,000 years old. This would fit well with the Pedra Furada site in Brazil that may be between 30,000 and 50,000 years old. Also, an older radiometric date has been recorded at Rye Patch, Nevada. Here bones of large mammals exhibit impact marks suggesting the use of blows for extraction of marrow; others exhibit negative flake scars formed by percussion, and all date at more than 20,000 years ago. It has been stated that 'The numerous possible bone cores are not unequivocal, but it is impossible to eliminate humans as a factor.' Michael Lemonick, writing in *Time* magazine, states that it is increasingly likely that the earliest American history will have to be rewritten."

"To take it one giant step further for humankind," Earth Walks boomed, "George Carter proposes that humans arrived on the North American continent even earlier, about 100,000 years ago, based on his work on a site in the Mojave desert; that date was agreed to by Louis Leakey. Carter has suggested multiple origins

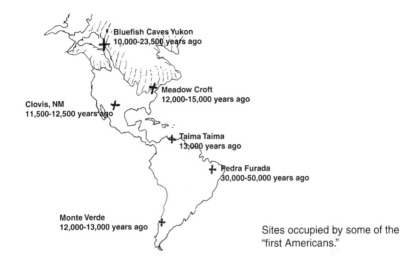

Bluefish Caves Yukon
10,000-23,500 years ago

Meadow Croft
12,000-15,000 years ago

Clovis, NM
11,500-12,500 years ago

Taima Taima
13,000 years ago

Pedra Furada
30,000-50,000 years ago

Monte Verde
12,000-13,000 years ago

Sites occupied by some of the "first Americans."

for the American Indians. Looking at the complex race picture in Indian America, he calls the idea that one pure race entered the Americas and subsequently diverged highly improbable. American Indian linguistic diversity certainly supports this view.

"It may be that we should take the Indians' accounts of their origins more seriously. Is it too improbable to suggest," Earth Walks asked, looking out over the audience, "that science may ultimately come to partly confirm the Indians' point of view on their origin as a people?"

"Time may tell," I responded. "And speaking of time, it's moving right along, as did the woolly mammoths and mastodons who followed the spread of the tundra and spruce forests northward. But they weren't alone. Along with them came other large creatures called mega-fauna, creatures that today would be the stuff of nightmares and tabloid newspapers. There were bear-size beavers, twenty-foot-long ground sloths, and roving packs of hungry dire wolves. Also, new to the spruce forest, were elk, caribou, and deer. There was game aplenty, and the ancestors of Peetwaniquot may really have believed that life couldn't get any better than this.

"Until something wicked their way came," Earth Walks said as I handed him the slide-changer and took my seat. "Over a period of about 2,000 years, from 12,000 to 10,000 years ago, not only the American elephants, but also 35–40 other species, were completely snuffed out. This wave of extinctions is one of the most puzzling events in ecological and geological history. The reason behind these extinctions has been debated for decades, and to date there is no final answer.

"Also, mastodons, mammoths, and their other large friends became extinct at the same time the northern forest was undergoing amazing changes. A warming climate saw spruce trees go out of fashion and a mixed forest of red pine and birch, along with jack pine and balsam fir, move in to take their place. Studies of pollen found in lake-bed sediments show that this change occurred about 10,000 years ago over a large region that stretched from Minnesota to Nova Scotia.

"As with Lake Agassiz and River Warren, the Paleo-Indians were around to witness this change, and possibly to have participated in it. They may have burned the spruce forests to drive animals into marshes, lakes, open ground, or wherever else it was easy and convenient to kill them. They may have seen this as a smart alternative to wandering through dark woods, looking for mastodon or elk, and meeting up with a party of hungry dire wolves. If the early Indians did burn the spruce forests, then they picked on a species of tree that hates fire and will not repopulate an area after it is burned. But pine, along with prairie grasses, regenerate after fires; their very existence depends on it. At the present time there is not a lot of evidence to support this scorched-earth idea, but it is certainly not out of the realm of possibility.

"Another popular theory attributes the animal extinctions to direct hunting and killing by humans. Human populations were expanding at this time, and more people require more food. Large, slow-moving animals, like the giant ground sloth, may have been easy targets for Indian hunters. Arthur Jelinek notes that the sudden appearance of humans would have had a formidable potential for disruption in an environment already under the stress of a warming climate.

"Such human-directed extinctions have happened elsewhere in the world. European sailors, in the fifteen and sixteen hundreds, made it a habit to stop at the island of Mauritius, in the Indian Ocean, to replenish their ships' meat supply. They did this by killing dodo birds, a large, flightless relative of the pigeon. Portuguese sailors brought pigs and monkeys to the island in the fifteen hundreds, and these animals destroyed the dodos' eggs and ate the young. The combination of humans, pigs, and monkeys sent the dodo down the highway of extinction, and by 1680 there wasn't a single one left on the island. Moas, a flightless bird that would tower above the tallest human, suffered a similar fate. When the Maoris arrived in New Zealand around A.D. 1300 they found twenty-seven species of moa, with a population density of eight hundred to one thousand per acre; shortly after the Maoris' arrival, the moas were nothing but bones in firepits. The Irish elk, with its twelve-foot antler span, was killed off in such great numbers that Irish monks reported fences made of their bones. In more recent times it has been the passenger pigeon, Eskimo curlew, Labrador duck, and Carolina parakeet brought to an abrupt and unhappy end due to human hunting, and finally there is the often-told story of the buffalo hunters and the near extinction of the American bison.

"From the historic past to the rain forest of the present, humans have sent many species on a one-way trip to oblivion. But this does not include the late Paleo- and historic Indians. These people were not in the business of extinctions; their religions had strict ecological codes that would condemn such events as the ultimate sin against Mother Earth. These people, then as now, saw their role in nature as one of reciprocal obligation, and this obligation included all living creatures. Certainly the continent that Columbus, Smith, Boone, Lewis, and Clark saw and explored was not a continent laid waste and despoiled, but it was, for them, a new world, an Eden rich in biological diversity. Could then the distant ancestors of these native people have looked upon the land and its life so differently? From what we know about them, probably not—unless they had a very good reason.

"In searching for such a reason it turns out many modern Indian societies have myths about mega-fauna. For the Ojibwe, their ancestral hero, Nanaboujou, born shortly after the earth's creation, pursues and battles giant animals that threaten human society. In myths he battles a giant beaver, driving it from the land, and finally kills it after a ferocious battle in what is now the St. Mary's River. He also pursues a giant otter and chases a giant moose, which he eventually kills. When he kills the moose its blood spills into a lake, turning the water red. The name Red Lake, a large lake in Ontario, endures to this day. All of these animals are large versions of animals alive today, animals that pose little threat to modern humans.

"So what then do these myths signify, if anything other than just tall tales? There is no question that post–ice age mega-fauna suffered population loss and mass extinctions; what is also important to remember is that humans of this time were themselves in danger and could have gone the way of the giant ground sloth or the dire wolf.

"Do these stories then reflect a time when these large creatures, struggling to survive in a rapidly changing and somewhat hostile environment, actually were a menace, or were seen as a menace, to humans? Did humans react by attempting to rid the world of what they perceived to be dangerous creatures? If so, what may be suggested by the Ojibwe myths is that Nanaboujou, who is the embodiment of early ancestral peoples, went about making the world a safer place for humans. What survived the mega-fauna extinctions were smaller, swifter, more adaptable animals: deer, moose, beavers, foxes, wolves, elk, and bison—animals that could coexist in a balance with, and were not seen as a menace to, humans.

"Finally, there was the warming climate. James Hester points out that there were widespread vegetational changes in ecological niches. This local change in food sources, coupled with the climatic extremes of hotter and drier summers and cold, bitter winters, may have had important effects on the breeding cycles and other critical aspects of survival for many species. Inflexible breeding habits and the inability to migrate to better habitats would bring pressure on populations of ice-age mammals and give them less time to adapt to climatic change. Harriet Kupferer concludes, and here I quote:

> *The answer to the question of extinction is not a simple either or. The most fruitful conclusion sees the stress upon animals by the changing environment, taking place slowly over several thousand years, combined with pressure on the herds from paleo Indian hunters. Neither humans alone or the ecology alone destroyed them.*

"In the final analysis, whatever the cause or causes, the mega-fauna came and went because they could not adapt and change as the land changed around them. Native Americans, on the other hand, survived and grew.

"And time passed.

"And over that time the Lake Agassiz people settled into a yearly cycle. They spent winter in sheltered valleys and woods; spring at the quarries, mining such rocks as flint, chert, rhyolite, and quartzite; summer at the lakeside; and autumn on the hillsides and in the woods, gathering plants, nuts, and whatever else for the coming winter. During this time, the planet continued to warm up, and around 8,000 years ago, it reached a post-glacial high that lasted for about 1,000 years. During this hot time three important geological and environmental events occurred."

"The first," I explained, taking Earth Walks' place at the lectern, "was the end of the Laurentide Ice Sheet. It had been hanging about in Hudson Bay hoping for a cold spell, but with increased warmth and decreased snowfall, it saw the writing on the ice and melted away. This caused a drastic increase in sea levels, and coastal areas became flooded.

"The second saw the rivers that ran through the land reach middle age and begin to tire out. Rivers had worked hard over the years to deepen their channels and erode the land away. Now they flowed slower, eroded less, and the land they traveled over slowly became flatter and flatter; in geological terms stream gradients decreased.

"Flowing at a slower rate meant the rivers could no longer carry all the sediment as before, and so this material settled out of the water like sand in a bathtub. Over the years sediment, in the form of silt, sand, and gravel, accumulated on the river bottom until, one bright summer day, brand new bars of gravel, or small sandy islands, appeared in the middle of what had been a free-flowing river. The river, finding it impossible to ignore these obstacles, was forced to flow around them. In this way rivers started to wander or meander, and rivers like the Mississippi, Missouri, and Ohio developed more bends, loops, and curves than the world's largest roller coaster.

"In the spring and early summer, with winter runoff and spring rains, the rivers flooded, and fine sediment was deposited across the rivers' floodplains. With much of the floodplain made up of clay-rich glacial till, the floodwaters were slow to drain away, and so new marshes, wetlands, and lakes formed along river margins. Animals and birds found this much to their liking, as did fish and shellfish. The Lake Agassiz people, and their relatives in other parts of the country, knew a good thing when it came along, so they made a permanent move to the river valleys. These first floodplain dwellers were apparently more observant than some of us more modern types, for they built their villages on the high ground above the floodplain, leaving the floodplain itself for animals, crops, and floods.

"Third, it was not only warmer but also drier. Drier times brought yet another change in the forests. The pine and birch that had replaced spruce of the wetlands were themselves replaced by a deciduous forest made up of oak, maple, elm, and basswood. This forest was not long for the north woods, for with the dryness and heat it was quickly replaced by prairie grasses. At its maximum, about 7,000 years ago, the prairie-forest boundary reached almost to Lake Superior, about eighty miles north of its present position. With the coming of the prairie so came the buffalo, and while the buffalo roamed on the plains, raccoons, turkeys, elk, and a great variety of deer romped through the forest."

Changing to a slide that showed a dugout canoe, I sat down.

"The Lake Agassiz people," Earth Walks said, "by moving to the rivers, adapted to the climatic, biological, and geological changes taking place in their environment. These river dwellers are referred to as people of the Eastern Archaic tradition by archeologists.

"When the Lake Agassiz people made the big move to the rivers, they built semipermanent settlements, and went into the boat-building business. Their boats were similar to the ones their ancestors used to paddle around Lake Agassiz, dugout canoes made out of hollowed-out logs. However, they made much more extensive use of their boats then their ancestors ever did. They used the slow, meandering, and less dangerous rivers for hunting, for exploration, and, most important, for travel so they could trade with other bands.

"River trading was a great change in their way of doing things. They were now able to set up trade networks that they could maintain and lengthen by the use of boats. Such networks brought a much greater variety of material goods to the Indians than they ever had before. There were seashells from the Gulf of Mexico; copper and agates from the Great Lakes region; silver and galena (lead sulfide) from what is now Missouri, Tennessee, and Illinois; obsidian from Yellowstone; and hematite (red ochre) and a great variety of flint from all over the land. These goods were used for tools, weapons, decorations, art, and ornaments.

"Many of these river traders were also self-trained botanists, and they were good at it, for their lives depended on it. They learned all the varieties of plants that grew in their territories, which ones to eat, make drinks out of, rub on, or stay well away from.

"And just how did they figure this out?" Earth Walks asked. "For figure it out they did, and to such an extent that by the time Europeans arrived, Indians used every potential medicinal plant, and every potential food plant in North America. Just as amazing is that the Indians often used these plants for the same purposes they are used today.

"The explanations Indians give is that a gifted individual would have a vision that directed the dreamer to the plant, and then she or he would be instructed in its use. Dozens of plant origins are explained in this manner by Indian tribes that have different languages, traditions, and plants."

"Visions and dreams are not the stuff of modern science," I told the audience as I stood beside my chair. "But is it possible a people could be so attuned to their natural environment that their instincts and senses, particularly those of smell and taste, could direct them to the right plants, and provide them with educated guesses on how to use these plants? This explanation, for me, seems far more savory than some of the other possibilities, such as trial and fatal error, large families to absorb mistakes, or the domestication of dogs for the purpose of chief tasters."

"Speaking of dogs," Earth Walks continued, "unlike many other seminomadic people of this time, the river people had no domestic animals other than the dog. Other seminomadic cultures in the world had domesticated such animals as chickens, pigs, llamas, sheep, horses, and camels. Most likely the river people had no need for carriers of burdens or tame food sources. They had boats for transportation and more small game and fish then they knew what to do with.

Plant visions.

"It has been suggested that the absence of large domesticated animals is one of the reasons Indian populations not only increased fairly rapidly, but also made for a hale and hardy people. Since there were no large herds of animals hanging around the village, there were no good hosts for nasty viruses, such as flu, and no camel spit or pig droppings to spread the viruses around, as often occurred in other cultures around the world. Smallpox, cowpox, trichinosis, and many other old world maladies are linked to farm animals and unsanitary conditions.

"Finally, the river people did not need domesticated large-animal protein sources because just over the horizon and through the woods lived the ultimate in culinary dining—the buffalo. The river people loved buffalo, and they were bison hunters par excellence.

"And now I think you have been sitting for long enough—it's time for a well-earned coffee break. My friend and colleague will be glad to answer questions, and I will be available after the lecture. However, at this moment I must excuse myself, for I have a little something to attend to."

PART TWO

THE NOTHING

Returning to the lecture hall after the coffee break, the audience was surprised to see how the stage had been transformed from bare boards to the set of a Broadway play.

A movie screen had been set up at the back of the stage, near the middle, and it was flanked by two large canvas panels, each of which showed a wide, open grassland under a bright blue sky. Leading away from the screen, on either side, were pillars made of what appeared to be boulders of a pink and grey rock, stacked one on top of the other to form columns four to five feet tall. These were set some five feet apart, and there were three on each side of the screen. Off to one side, close to the exit, was a solitary gray boulder.

When everyone was seated, Earth Walks, a buffalo robe draped around his shoulders, his hair again long and loose, moved into the center of the stage.

"Bison hunters par excellence," he said, repeating the last thing he had said before the coffee break. "Try and imagine how Native Americans went about getting a small part of a large buffalo herd to commit suicide by stampeding over a rather tall cliff. This obviously had disastrous results for the buffalo, but I think it clearly showed who had adapted better to the land!

"Getting the buffalo to commit mass suicide took knowledge of the land, detailed planning, perfect timing, and luck. To see how this may have been done, travel back with me some 3,500 years to what is now southwestern Minnesota. There we will join twelve-year-old Kwe Mnimiza Tinta as she secretly watches a sacred buffalo drive.

EARTH WALKS: Summer, 3,500 years ago, in what is now southwestern Minnesota.

> (A young girl, a buffalo robe on her shoulders, crawls onto the stage and wiggles behind the grey boulder. At the same time six boys, dressed in traditional Indian garb, come on stage from the other side. Each positioned himself behind one of the rock pillars.)

In her dream the buffalo were pouring over the red cliff like logs over a waterfall, when Kwe awoke. She was not sure if it was the sun on her face or the noise that had brought her back, but she had fallen asleep when she had promised herself she would not. Scrambling to her knees, she wiggled out of her warm, brown robe like a snake out of its old skin. Crawling forward, she peered around the large boulder.

> (The girl skillfully, and enthusiastically, enacts everything Earth Walks says and continues to do so throughout the program.)

Kwe's dark eyes grew as large as the copper discs she wore around her neck, and she was sure her heart beat louder than the thundering of all the heavy hoofs headed her way.

(The lights dim and the movie screen comes alive with scores of running buffalo. The noise of their hoofs thunders around the auditorium, then slowly dies into the background as Earth Walks continues speaking.)

Spread in front of Kwe was a wondrous sight. A rich, snorting river of brown that flowed across the grassy plain like shadows before the rising sun.

KWE: A river, with hundreds of currents, and it is the job of the drivers to keep each current headed in the same direction. And this is not easy, for the currents are large, lumbering bison that have been cut from the main herd, and that are now being skillfully driven over the prairie toward the nothing.

See! *(In a high, trembling voice edged with excitement.)* There! Through the rising dust, the small figures, drivers with their torches held high, dark smoke curling up and away like mist blown across a lake by a strong wind.

If the herd tries to turn, the drivers will use the torches to set the grass on fire and keep the frightened animals moving ahead of them.

EARTH WALKS: With a loud snort, the first of the beasts entered the run, and, like a formation of geese flying south, the rest were right behind. The run was a natural dip in the flatness of the vast plain, a long antler-shaped corridor that was widest where the animals entered, and narrowest where it met what Kwe called "the nothing." Along both sides of the run were columns of pink rock made by piling several large boulders on top of one another.

The columns, which looked like bloody ribs stuck into the earth, each hid one or more of Kwe's people. Turners, Kwe had named them, and there were more turners than she had teeth. Once behind the pink rocks the turners had to remain as still as a blade of grass on a windless day. If a turner made a noise or moved at the wrong time, the animals could panic and stampede every which way. They could even turn completely around, and that would be disaster. There were stories of this happening, and of the drivers who got caught in the middle of hundreds of crashing hoofs.

Once the animals were in the run, everything depended on the coordination and timing of the turners. If the lead animals tried to turn aside and break out of the run, they had to be stopped. The turners had to leap out at just the right moment; too soon and the herd could be completely spooked, too late and the turner could be trampled to death, and the herd lost to the prairie grass. It had to be just as the animal dropped its shoulder, the signal it was about to turn. Leaping out— *(Three of the boys jumped from behind the pillars ringing bells and beating drums they hopped up and down and screamed as though their very lives depended on it. Earth Walks shouted over the noise.)*

They would do anything and everything to keep the bison heading straight down the run. If the turners failed, they could lose their lives, which would be easy compared to the disgrace for the rest of their families. Today the turners, Kwe's clansmen, were quick of eye and full of powerful medicine, for they kept the brown river rolling straight for the nothing."

KWE: *(Standing up and facing the audience.)* I call it the nothing, because it is the place where the run ends, the sky begins, and the bison become one with the spirit world. The nothing is the top of a high cliff with a long, steep drop to the ground below.

(Pauses for a moment to look down at her feet. Lifting her head, a look of pure panic spread across her face.) I forgot! In the excitement I forgot where I am and what will happen if I am seen. If a turner or, far worse, the Holy Man himself happens to choose this moment to look my way I will be lost. Truly lost, for it is forbidden for me, a girl, to be here, to be watching the sacred drive. *(Quickly ducking down, she harshly scolded herself.)* I should bite my tongue off. *(She giggles and shakes her head. Her body shaking like a leaf, she slowly peeked around the boulder and continued to stare at the buffalo on the movie screen.)*

EARTH WALKS: A tremendous series of squeals and bellows came from the lead animals. With white froth flying from their mouths, the bison dug their hoofs deep into the grass, desperately trying to slow their forward rush. Dropping both shoulders, heads almost touching the ground, they bellowed madly, dust clouds rising around them as they skidded forward like otters on ice.

KWE: *(From behind the rock.)* They can bellow and dig as much as they want, but it is far too little and much too late. The great beasts at the front, the ones who have seen and smelled the nothing, are pushed forward by the moving mass behind them, just like dark clouds chased across the sky by a fast-moving wind.

I can hear the frightened bellows of the lead animals, see them hanging at the cliff edge for a lightening bug flicker. Then, in the blink of an eye, they are gone. Great logs, giant trees swept over a waterfall by a fast-moving current, crashing down, slamming one into the other, the pile of death growing faster than a termite mound in the sand. It is just like my dream, and that is surely a good sign."

EARTH WALKS: When the Holy Man decided enough animals had been driven over the nothing he raised his medicine stick high into the air; a painted buffalo skull, tied to the end, shone blood red in the morning sun. *(Picks up a long wooden pole; a buffalo skull, painted bright red, was attached to one end. Raising it high into the air, he moves it back and forth.)* Waving it back and forth, the holy man signaled the end of the drive. The people behind the tall ribs gave forth a great yell followed by several long whoops.

> *(The six boys let out several long whoops and then they sing what Earth Walks calls the sacred buffalo song and move off the stage.)*

The hunt was over, and Kwe had seen the whole thing. She had actually done it, and never had she felt so good in all her twelve years.

Buffalo swept over the cliff like giant trees over a waterfall.

KWE: *(Grabbing her robe, wiggling out from behind the rock, and walking to the edge of the stage.)* The Holy Man said this was a good day to hunt. All the signs were right, and my dream has proven that. Still, it would all end badly if they caught me out here in the prairie grass. But I have plenty of time before being missed. It isn't every day the tribes come together for a hunt; it happens only once or twice a year. With all the talking and bragging and storytelling, no one will miss me, especially when it is my band's honor to be the turners. I have lots of time to get back to my lodge.

EARTH WALKS: *(Coming to stand next to Kwe.)* Kwe needed to be there before the camp moved to the kill site. It was tradition for the camp to wait for the runner to come, to tell them that the hunt was a success and that it was time to go. The entire camp would then move. As a group the people would go to the cliff. Songs would be sung, thanking the buffalo for their gift of life, and the tongues would be cut out and shared while the Holy Man and turners and drivers sang the buffalo songs. Then the cutting, drying, and feasting would start, and this could last for several days.

KWE: *(Looking up at Earth Walks.)* I can taste the sweetness of the warm tongue, and smell the sharp odor of the animals' hot insides.

(Earth Walks holds out his hand to her. She clasps it firmly, and the two of them bow to the audience.)

The applause was louder and longer than that of the first night, but as on the first evening Earth Walks held up his hand to stop it. When the crowd was quiet he introduced his young actress, and then the actors, thanking them all for a wonderful performance. To the audience he said, "the idea of stampeding buffalo over tall cliffs represents one of the best land and natural resource management plans I have ever heard of. In the matter of a few days, which includes butchering and fire- or sun-drying tons of meat, a band could process enough meat to last for an entire year. Along with all this meat they also had hides, fur, and bones, which they utilized for most of the other things they needed. So the next time you see a buffalo, instead of looking at it as a shaggy brown beast that smells bad, look at it as an early Native American might—a one-stop shopping plaza. A place where you could get groceries, hardware, furniture, clothing, and a hundred other necessities that allowed you to live well with Mother Earth.

Earth Walks then inclined his head slightly toward the audience. "There will be a short break so we can clear the stage. Then the program will resume."

PART THREE

FOOD TO GO

The stage clear and everyone seated, Earth Walks, minus the buffalo robe, started the final part of that evening's program.

"It was tradition for North American Indians to drive or stampede bison over cliffs, or to herd them into box canyons and shallow water, making it easier to slaughter them. They had been mass-hunting the buffalo for more than 8,000 years, and this event, in the turning of the seasons, was a most important communal activity. Not only did it give the different tribes the opportunity to trade and exchange information, but it also allowed them to tell stories, meet new people, arrange marriages, and have communal ceremonies.

"Some 5,000 to 6,000 years ago the Canning Site, located in the Red River Valley floodplain in Norman County, northwestern Minnesota, was one such place. Hundreds of fragments of buffalo skeletons have been found there, and most of these fragments are parts of legs or ribs. This strongly suggests that the bison were dispatched elsewhere, and only the choicest cuts brought to the Canning Site for further processing, much like the movement of meat today from slaughterhouse to butcher shop. Artifacts at this ancient butcher shop suggest meat was cut and scraped off the bones, possibly to be sun-dried and made into that all-time gourmet treat: pemmican."

Saying this, he stepped to the edge of the stage and inclined his head toward one of the exit doors. Immediately the doors opened, and his young actors and actress appeared, carrying large trays full of long, brown strips of meat. Moving amongst the audience, they gave each person a piece.

"Pemmican, in case you want to whip up a batch for the holidays, is a mixture of dried meat, a piece of which you now hold in your hand, along with fat and berries. Variations existed in almost all parts of Indian America, wherever hunting took place. In times before refrigeration it was a foolproof method of preserving meat, and it surpassed almost all other systems. It was a staple on the plains, and later it was the fuel for the voyageurs of the fur trade and actually became a trade item itself. Variations of it are still being made today as trail food, and Apollo astronauts took a modern version to the moon.

"The importance of pemmican in the history of North America should not be underestimated. It is the American Indians' greatest contribution to the food systems of the world. Indian societies did not use fresh meat as a staple, they used pemmican.

"Cliff or river drives were difficult enterprises to organize and carry out successfully. Only a small portion of a buffalo herd, which would number in the hundreds of thousands, could be directed to a suicidal charge. Such communal hunts, before the days of the horse and rifle, were biannual events. Between hunts, pemmican was the daily bread, and there was probably no greater innovation in the human food system, with the exception of the invention of agriculture.

"To make pemmican, Indians cut buffalo meat from the bone into slabs four feet long and one inch thick. The fresh meat was dried over a fire or in the sun for several days until it was the consistency of wood. These dried slabs were then placed on a buffalo hide, staked down to the prairie, and pounded with wooden mauls until the meat was reduced to pulp. The pulp was then placed in a rawhide bag. Into this was poured tallow, which came from the slaughtered buffalo and had been prepared separately in rawhide pots. Dried berries, if available, were mixed with this; then the bag was sewn shut with sinew. Prepared in this way, the mixture hardened, and could be kept for years.

"'Harder than cement' could be changed to 'harder than pemmican.' The mixture dried so hard an axe was needed to break a portion free. The chopped-off chunk would be boiled with berries, groundnuts, lily roots, or vegetables; it reconstituted itself so well in this boiling broth that only a small amount was needed to feed a family of four. In other words, it could last a long, long time. And we get sick of leftovers after one night," he observed with a smile.

"I would like to finish this session by taking Native Americans into what are called historical times," Earth Walks said. "To do this, all I have to say is when you're hot, you're hot, and when you're not, you're not. And about 4,000 years ago the earth became a not. The planet became colder, and climatic conditions moved closer to those we have today. The advance of the prairie came to a grinding halt, and the grasses retreated to the south and west. The northward march of what was to become the great white pine forest began, and the meadowlands that had spread across the dried-out bottom of Lake Agassiz were converted into what is today called the Great Minnesota Bog.

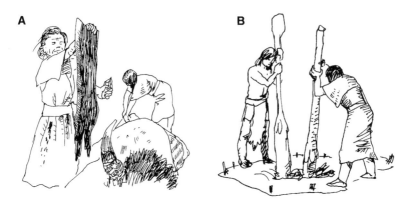

To make pemmican, (A) buffalo meat was cut from the bone into slabs four feet long and one inch thick. (B) The sun-dried slabs were placed onto a buffalo hide and pounded until reduced to a pulp.

"With the change from dry, flower-filled meadowlands to fly-infested bogs, came a change in the river people. Around 3,000 years ago they founded a new tradition, mound building, took up new crafts, like pottery making, and started a new pastime, which has only now reached its cultural maximum: gambling. With these three additions our river people, or what archaeologists call the people of the Eastern Archaic tradition, moved on to become people of the Woodland tradition.

"Three brand-new innovations," Earth Walks announced as he rolled a pair of large dice across the stage floor. "Innovations that included spending the long winter nights chanting moccasin game songs as the bone or antler dice rolled away with their fortunes in agates or copper. Despite these cultural additions, the river people's basic way of life remained much as it had been. They continued to live on the river, they collected and cultivated numerous plants, they were heavy into fishing and boating, and they continued the tradition of seasonal migrations to take advantage of the natural foods found in their territory, and the semiannual buffalo hunt.

"The new phenomenon in town was mound building. It is not known where this idea came from. But whether the river people got it from those they traded with, or from newcomers into their territory, or they thought of it themselves, it was like lightning striking dry prairie grass. Earth mounds sprang up across the countryside faster than potholes in Minnesota roads at winter's end. The mounds appear to have had two basic functions; one was to serve as a burial chamber, and the second was to serve as a platform on which buildings could be constructed.

"It is estimated that there are more than fifty thousand mounds in the Pays d'en Haut. The mounds came in all sizes and shapes, though the two most popular seem to have been circular and domelike, or long and narrow. The people of this time also constructed effigy mounds, mounds that were built in the shape of different animals such as birds, fish, and snakes. The most famous of the effigy mounds is the Great Serpent Mound near Hillsboro, Ohio. This was built by a people called the Adena more than 1,200 years ago. It has been suggested that the effigy mounds were built in the shape of the different animals the people saw outlined in the stars overhead. Whether the purpose in doing this was to honor the star creatures, or to help the deceased on their journey to the stars, or both, is not known.

"The largest preserved mound constructed by the river people is Grand Mound on the Rainy River; it is one hundred feet in diameter, and it stands more than forty feet high. The biggest mounds, however, were constructed by farmers who started to move into the southern part of the Pays d'en Haut about 1,200 years ago. These Native Americans are called people of the Mississippian tradition by archeologists, and they cleared off wooded areas, mostly on the rich soil of floodplains, or they dug fields in areas covered by thick blankets of glacial loess. In these fields they grew such things as maize, beans, squash, sunflowers, and tobacco, and in their spare time they built great mounds.

"The greatest mound they built was the so-called mega-mound of paleo-America, located at the great Mississippian trading center of Cahokia, Illinois. This mega-mound, called Monks Mound, was ten stories high, with a base larger than that of the Great Pyramid of Egypt. It is believed this stairway to heaven served as a platform for a building that housed the autocratic ruling family of Cahokia.

"Excellent examples of both circular and elongate mounds can be seen at Tamarack National Wildlife Refuge near Detroit Lakes, Minnesota. In regard to these, and the glacial geology I have learned the last few weeks, it is interesting to note that in pre-mound-building days, the river people buried their dead in glacial beaches and eskers, which are long and narrow, and in kames, which are round and domelike. Leads one to speculate about the connection between glacial landforms and star creatures. Mound building couldn't be explained as easily as that—or could it?

"Anyway, back to what I know a little something about. When the first Europeans arrived in mound country they could not, would not, believe their eyes, and they certainly didn't believe the native peoples they came in contact with were at all capable of building such imposing structures. So the Europeans invented fantastic theories to explain the mounds. These ranged from the silly, such as attributing them to the lost tribes of Israel, through the improbable, such as roving Vikings. However, it was not until the twentieth century that we reached the downright ridiculous, that space aliens built them, I suppose with antigravity machines in the holo-deck. Few white people took the time to ask the Indians if they knew who the great constructors were, and the few that did were told the simple truth: The peoples who were here before us, ancestors of Native Americans when they lived in large towns and whose people were as numerous as trees in the forest."

And that concluded the third lecture.

CHAPTER FOUR

EARTH ROOTS

PART ONE

INTRODUCTION

"Let me begin by thanking you," Earth Walks said, looking relaxed and confident standing at the edge of the stage.

"Giving up a nice Saturday afternoon to come and listen to the likes of me—that's dedication or interest or masochism. And whichever, I much appreciate it.

"As you know," he continued, smiling at the audience, "this is an extra lecture in the series, one I specifically asked to have added. I thought it important that you know something about the Native Americans who today call a good part of the Pays d'en Haut home, and who have done so for more than six hundred years.

"Since the theme for the lecture series is living with the land, living with geology, that will be the focus of my remarks concerning these people.

"To begin, when the Europeans arrived in the Pays d'en Haut, around A.D. 1600, or the beginning of the so-called historical period, they found the native people divided by regional differences," Earth Walks explained, glancing down at the single sheet of paper that represented his notes. "The people in the north were mostly hunters and gatherers in the Woodland tradition—what we have been calling the river people. Those living in the south and east were dominantly Mississippian farmers, and the people living on the high plains were much like the northern people, seminomadic hunters and gatherers, though they spent much of their time hunting one commodity: the buffalo. This is the way it had been and the way it was in the Pays d'en Haut—though not for much longer.

"Great political and cultural changes were taking place, and these would obscure, and in some cases obliterate, these regional differences. The fur trade was being established in the Lake Superior region, the horse had recently been introduced to the northern plains, new Indian peoples were arriving in the Pays d'en Haut from the east, and the influence of the Mississippian culture in the region was on the ebb. Starting about 900 or 1,000 years ago, and continuing through the time these events were taking place, there arose, through the consolidation and blending together of many of the different peoples living in the Pays d'en Haut, two new Indian entities, the Dakota, or Sioux, and the Ojibwe, or Anishinabe.

"Both of these groups maintained the basic belief that they were directly connected to the land; they were one with Mother Earth. This, for them, meant their actions and thoughts, good or bad, had a definite impact on the land and its animals and, through these, on their existence as a people."

PART TWO

PEOPLE OF THE ROCKS

"The people known as the Dakota or Sioux had a very complex society," he stated, staring out over the audience. "It was based upon close kinship ties and reciprocal obligations. The Dakota taught children that everyone in the village was their kin, and names such as mother, father, uncle, aunt, niece, and nephew had far greater and more important applications than in

European societies. The Dakota term for 'father' would include cousins and uncles as well as the biological father. Often these uncles and cousins took more responsibility for teaching, training, and providing learning environments than did the biological father. The idea of an isolated nuclear family was as alien to the Dakota as living in skin teepees and eating buffalo tongue would be to us. For the Dakota, society provided a place for everyone.

"In exchange for this, an individual had strict obligations to the village. Everyone was expected to do their share of the work, and to perform their duties, according to gender and age. Everyone was also expected to share what they had, so the idea of charity was unknown.

"Religion governed many aspects of Dakota life. 'Wakan' was the word the Dakota used to encompass the all-powerful force in the universe, and their world is sometimes translated as 'the great mystery.' Wakan Tanka, the Great Spirit, was the creator of all life; and animals, plants, rocks, and rivers were imbued with powerful medicine that could be benevolent or threatening. Offerings and prayers could win favor from the natural world, and medicine lodges were societies of shamans that kept the world in balance. No enterprise was undertaken without prayers and/or sacrifice. Sharing with kinfolk kept the spirits of animals in a person's favor, and proper and respectful behavior to elders would be noted by the natural world. Everything one did in the human community had repercussions in the natural world; thus unusual behavior of animals, sudden geological events, or the appearance of rare weather phenomena would have positive or negative implications for the entire community. Being close to nature, for the Dakota, meant being in a constant religious state, strictly observing their responsibility to their relatives, especially Mother Earth.

"One of the most outward expressions of this responsibility was the smoking of the sacred pipe. Communal smoking confirmed the bond between family, tribe, Mother Earth, and the universe. The bowl of the pipe was made of red stone, commonly from the pipestone quarries in southwestern Minnesota, and was meant to

White Buffalo Woman offers the first
sacred pipe to the Lakota Sioux.

represent the earth. The wooden stem, which represented all growing things, was decorated with feathers and designs reflecting the owner's personal spirits and visions.

"According to legend, the first sacred pipe was given to the Lakota Sioux by White Buffalo Woman. In giving the pipe to a Lakota chief, she is supposed to have said, 'I am White Buffalo Woman. This is very sacred, and no impure man should ever be allowed to see it. With this, during the seasons to come, you will send your voices to Wakan Tanka.' White Buffalo Woman then took the pipe and, holding its stem toward the sky, said, 'With this sacred pipe you will walk upon the earth; for the earth is your grandmother and mother, and she is sacred. Carved into the red bowl is a buffalo calf, which represents all four-legged creatures. The stem is made of wood and represents everything that grows.'

"The woman then announced her departure, saying she would return one day. As she walked away from the gathered people, White Buffalo Woman first became a young red and brown buffalo calf, than a black buffalo. The buffalo bowed to each of the four sacred directions of the universe, then disappeared over the hill.

"Another sacred tradition, not only of the Sioux, but of many Indian people who lived in the Pays d'en Haut, was the sweat lodge ceremony, a symbolic rite of healing and purification.

"In Sioux tradition the diameter of the sweat lodge is determined by the height of the builder. He or she lies down north to south, then east to west, marking the places where the feet and head come to rest. The four marks represent the four virtues: respect, courage, generosity, and wisdom. A circle is drawn around the four marks, and sixteen willow branches, four for each mark, are collected and used for the lodge frame. The branches are placed so they form a dome outlining the drawn circle. The branches are then covered by skins or a heavy cloth.

"Rocks are heated in an outside fire, then quickly lifted out and placed inside the lodge. A pipe ceremony is conducted, and the participants sit in a circle around the rocks as sage burns. Prayers are sung and tobacco is offered to the hot rocks. The glowing stones slowly turn dark as water is splashed upon them. The pipe is passed around the circle, the heat builds, and healing steam fills the lodge.

"In the far west, the sweat lodge ceremony concludes with a quick dip in an icy stream or soak in a hot spring—ending rituals any Finn or Norwegian would recognize. Indeed, when Finnish immigrants arrived in the north country and built saunas on their homesteads, the Ojibwe called them *Mi do do jig,* or People of the Sweat Lodge.

"The Dakota legend of the origin of the sweat lodge begins with a young girl who lived with her five brothers beside a creek. Each day her brothers would go hunting, but soon something terrible began to happen. At the end of each day one brother would vanish, until the girl was left alone. In despair, she wanted to kill herself, so she swallowed a large, dark stone. Instead of killing her the stone grew inside her, and she eventually gave birth to a boy, whom she named Stone Child. Soon Stone Child began to hunt, much to the fear and sorrow of his mother.

"Once, on a hunting journey, Stone Child came across a tipi in the woods. Inside was an ugly old woman who sat in the middle of a circle made out of five bundles of rock. Stone Child suspected the old woman was a witch, so he lured her outside the tipi and killed and burned her. Then Stone Child heard soft voices from the rock bundles, voices that told him to build a lodge of poles

Stone Child, the witch, and the five bundles of rocks.

and hides. He was instructed to heat the five rock bundles in a fire and then to place them inside the lodge and pour water over them.

"He did as he was asked, and out of the steam emerged his five uncles. The family was reunited back at the mother's camp, and ever after the rocks and sweat lodge have been sacred to the Lakota.

"The Dakota people placed great religious importance in Mother Earth and the rocks of her body. There were the sacred stone pipes, the sacred rocks of the sweat lodge, and Wakan Tanka, the Great Mystery, whose spirit was the very first god, Inyan, or rock. These people knew that in the beginning there were only rocks, and therefore rocks were sacred spirits that needed to be recognized and honored."

PART THREE

PEOPLE OF GICHI GAMI

Momentarily pausing, Earth Walks sipped at a glass of water while glancing at his notes. Taking a deep breath he loudly cried, "Ojibwe, Chippewa, Anishinabe, Gichi Gami! The first three are the names commonly used for the native people who live in the Lake Superior region, and the last is their name for the great lake they call their home.

"The word 'Ojibwe' is used whenever discussing their culture or science, such as Ojibwe language or Ojibwe constellations. Chippewa is used on treaties, for political relationships, and in many formal institutions, such as the Minnesota Chippewa Tribe. Anishinabe is what the Ojibwe call themselves, one Ojibwe to another.

"The Ojibwe have always been autonomous bands, independent from one another; there was never a 'chief of the Chippewas,' as some historic Indians have been called. What bound these bands together was a system of clans, or grand families. There were some thirty different clans, and each had its own emblem or symbol, which was usually one of the animals with which the Ojibwe shared the land. The Ojibwe assigned characteristics to these clan animals based on their own interaction with them and their long observation of the animals' habits. Thus the bear clan was a warrior clan, a clan of strength and power, whereas the crane clan was one of leadership, based on the bird's loud, resounding call, to which everyone would listen. Each clan had a specific purpose in society; thus there were clans for healing, hunting, learning, leadership, and so on. A person inherited their clan type from their father, and it was strictly forbidden to marry someone else from your own clan, even if the person belonged to a different band.

"Another important part of Ojibwe tradition was the *Midewiwin*, or Grand Medicine Society. This intertribal society was like the American Medical Association, Royal Philosophical Society, and Smithsonian Society all rolled into one; its members were healers, herbalists, prophets, and scientists. There were Midewiwin members in other Algonquin nations, as well as every Ojibwe band. The medicine men and women of the society were intermediaries between the people, the land, and the spirit world, and were therefore called upon in times of crisis. The members of the society tried to interpret or explain natural phenomena, and they constantly watched the skies from great granite cliffs or basalt ridges. These astronomical sites were commonly marked by pictographs, which linked moral teachings, myths, star lore, and natural phenomena.

"The origins and traditions of the Grand Medicine Society are obscured by the mists of time. There was a time when the Ojibwe lived on the Atlantic Coast and a time when a great sickness walked amongst them and the people suffered terribly. Then came a sign, a great shining seashell that magically appeared in the western sky and beckoned them toward a new land—a place, it was foretold, where food grew on the water. This mystical seashell came to be called the Megis, and it became not only a symbol of healing for the society, but also the basis for a new religious creed. And so, as tradition claims, began the people's great migration to the west, following the Megis to a safe, new land, a land where food—wild rice—did indeed grow upon the water. This was a holy journey along the St. Lawrence River to Montreal, and then on to Sault Ste. Marie and Lapointe. This journey was much like Abraham's journey out of Mesopotamia, and as with Abraham, everyplace the Ojibwe stopped to rest later became a holy site.

"The Grand Medicine Society also had a set of moral teachings, which emphasized sharing, respecting the land and all the animals that lived upon it, and helping orphans and the elderly. It also set out rules of conduct for all to follow.

"These rules were based on the belief that humans were the weakest of God's creations, and this weakness required harvesting a part of nature to feed, clothe, and shelter themselves. Therefore, hunts and harvests not only were to be conducted with ceremony, honoring the sacrifice of that which was harvested, but were to be done with restraint. Abuse and waste would mean that, in the future, Mother Earth would withhold her gifts to the people.

"In Ojibwe law petty social infractions were reprimanded through ridicule. The Ojibwe punished serious violations with exile. The crimes that were considered serious were actions that threatened community survival, such as breaking into a winter cache or disobeying the rules of the rice harvest. The majority of these serious crimes were perceived as 'crimes against the environment': violations of the codes and taboos that ensured respect for the land and a plentiful food supply. Maude Kegg, an Ojibwe author who lived at Mille Lacs Lake, tells of a rice committee on Mille Lacs that punished two men for ricing too early; their canoes were smashed to pieces in front of the whole village.

"The Ojibwe, much like the Dakota, believed they were one with the land, physically and spiritually. For them the ultimate expression of this was the changing of the seasons. As the seasons changed, so did Ojibwe camps and villages, hunting and gathering activities, social obligations and overall lifestyles. It was this seasonal change that defined them as a people.

"As an example of this, there is the story of the two white anthropologists who came and studied the Ojibwe. One wrote that the Ojibwe were communal and intensely social, and lived in large towns. The other wrote that they were isolated, loners, and individualistic, caring little for their fellow man or woman. These two scholars used to debate and argue all the time, one never believing the other. As it turns out both were right, for one visited the Ojibwe in summer, the other in winter! To find the truth, these two scientists needed to be more like the people they were attempting to study—they needed to become people of the seasons."

PART FOUR

THE FOUR SEASONS

"In that regard I would like to end today's presentation by sharing with you a description of a typical Ojibwe year. This will, I hope, illustrate how the lives of these people revolved around the changing seasons, around the ebb and flow of Mother Earth."

As Earth Walks spoke, a young woman, wearing a soft white doeskin dress, rabbit fur fringing the cuffs and bottom, walked across the stage to stand beside him. "The description I am going to read," she said in a warm, soft voice, "was written by an Ojibwe woman, Bineshiikwe, many years ago."

As Earth Walks left the stage she opened a thin leatherbound book and began to read:

We Ojibwe women have much responsibility and many duties, for it is we who lead our people through the four seasons of our year.

The Ojibwe year begins with the death of the Wintermaker. When he falls from the sky, we know it is time to leave our winter camp and travel to the sugarbush village to be with other clan members and to make maple sugar.

We women are in charge of collecting the maple sap in birchbark troughs and boiling it down to sugar. We make either granulated sugar, by stirring the thick syrup with wooden paddles, or hard cake sugar, by pouring the syrup into carved wooden molds or birchbark cones. This sugar we use as a spice much more freely than the Americans use salt; we also use it to sweeten meats, fish, stews, and soups. In the summer we dissolve it in water for a cool, sweet drink, and it can be mixed with medicine for the children. Our family makes five hundred to eight hundred pounds in four to six weeks, and this is more than enough for the entire year.

Spring is also the time for the making of new canoes, our main form of transportation. It has been said the Dakota have their horses, the whites their trains, and the Ojibwe their birchbark canoes, the most efficient and reliable of the lot. Our birchbark canoes are easily portaged, float high on shallow streams, and travel easily through the rough waves of big lakes. Our canoes and knowledge of the waterways have enabled my people to prosper for hundreds of years; with our canoes we trade, hunt, fish, gather wild rice, explore the land, and travel from berry patch and beaver meadow to the places where the wild plants grow.

We make canoes in the spring and early summer, for this is the time when the birchbark is soft and pliable. Both men and women work together, and we teach our skills to the younger ones who assist us. Our family, like most, owns several canoes, all of different sizes, some built for speed, others to travel across big lakes, and yet others to carry much weight.

In the summer, which we call budding seeds time, we women plant and tend the gardens, which includes keeping pesky birds away. In these we grow potatoes, squash, pumpkins, corn, beans, and sunflowers. I also plant peas, which I very much like, and parsnips from seeds I get through trade.

Once the gardens are planted, most of us women and children take our canoes and go off to pick berries. We eat many of these fresh, but many more we boil down and sun-dry for winter use.

We also pick wild ginger, mint, and bearberries to use for seasonings. We add wintergreen, spruce moss, raspberries, and twigs of the wild cherry to our water for flavoring, and we eat the flowers of milkweed, roots of bulrushes, sap of the basswood and aspen, and moss of the white pine, along with delicacies like fiddleheads, marsh marigold leaves, and wood sorrel. To our soups and stews we add cornsilk and pumpkin blossoms, which thickens them and adds flavor.

We always have a great celebration and feast for the first fruits of the summer. During this feast we always ask Manidoo for health, long life, safety, and good harvesting.

We women also take part in the spring and fall fish runs. We use nets that we make out of nettle-stalk fiber. I am not one to brag much, but we women catch as many fish as the men with their hooks and lines, their traps, and their fierce-looking spears.

I must admit though that seeing the men spear fishing is a beautiful sight. They fish this way at night and by torchlight, suspending fire baskets from the bows of the canoes, which makes the water transparent to a great depth.

We eat some of the fish we catch fresh, but most we smoke and store for later use.

Summer, in the old days, was the time to make war. War, for us Indians, was always restricted in time and place. By mutual consent our wars had to fit between canoe-making time and garden harvest. I imagine Americans and Europeans will think this very peculiar, for it contradicts their talk of us as fierce warrior societies. Our wars would stop when it was time to gather in the pumpkins! This is how it was and how it had been for as long as the wolf has hunted in the north woods.

Spear fishing by torchlight.

My grandfather, when a young man, met the most renowned Ojibwe warrior on Lake Superior. Wattab was his name, a dignified man of middle age, and a veteran of many battles with the Dakota. My grandfather saw he wore seven feathers in his hair, all dyed red. My grandfather asked him if this meant he had killed seven Dakota. That is a good number, Wattab replied, but I killed only four really; the others were shot by my friends.

Seven? Four? This is surely not war in the European or American sense. Why, by father's time Napoleon had changed the rules of war, and the Americas were busy slaughtering thousands in what they call their Civil War. Death such as this bears no comparison to the modest little fights between us Ojibwe and the Dakota.

No matter, for war is still war, and people die. Hearts are broken, there is much sadness, and there are families with no one to hunt for them. Our wars were not games, but I think there could be a better word for our conflicts than the term 'war.'

Enough said about such things. I will tell now about our wonderful fall season, which belongs to the lakes and the Manomin (good berry), or wild rice.

I once heard someone say that we Indians who harvest wild rice are as well off as most gardeners at harvest time, for we have a crop that will last through the winter. It is also true that we Indians who eat wild rice are noted for our good health.

Our family always goes to the same lake and the same marsh to harvest the rice. Our part of the lake is marked with our stakes and clan totem. We women go to the marsh in late August or early September to tie the standing stalks into bunches, making it harder for the birds to get at the ripening grain, and also making it easier to harvest.

Harvest time brings with it a first fruits ceremony, and all the ricers on the lake take part. We hold this to thank the spirits who inhabit the plants, the water, and the earth. On the first trip to our rice marsh we always leave a tobacco offering on the water and thank the spirits for their great gift to our people.

As it does for many things, it takes two to gather the rice. One person guides the canoe with a long pole, but sometimes the rice plants are so thick and so close the canoe has to be pulled through them, one rice clump at a time. The second person, usually one of us women, bends the tied seedheads over the canoe, and with wooden sticks knocks off the rice kernels onto woven mats that line the canoe's bottom. This is done about as fast as the loon swims, so there is a constant sprinkle of seeds onto the bottom of the canoe. Much rice is lost to the water spirit. But that is as it should be, for it gives food to the geese and ducks, and seeds the marsh for next year's crop.

It is said that it was 1,000 years ago when we Ojibwe discovered how to harvest and preserve rice. The kernels are roasted and then threshed, the hard hulls being winnowed away by the blowing wind. If properly stored, such rice will keep for many seasons. I think our Indian diet of wild rice, maple sugar, meat, fish, vegetables, and herbs is as healthy as any in North America.

I love the wild rice harvest, the smell of the roasting rice in the crisp fall air, the music of the flailing sticks, and the dances and songs of the powwow. These are happy, satisfying times, but also sad times. Sad because the rice camp marks the end of the time our people will live together in large groups until the following spring. The days slowly grow shorter and colder; winter, our harshest season, our time for family groups and small camps, is fast approaching.

Like the sugarbush and ricing camps, our winter camp is always in the same part of the forest, for this is our family hunting and trapping area, and it is passed down from father to son. Winter camp is a time of isolation, storytelling, hunting and trapping, and sewing. During these

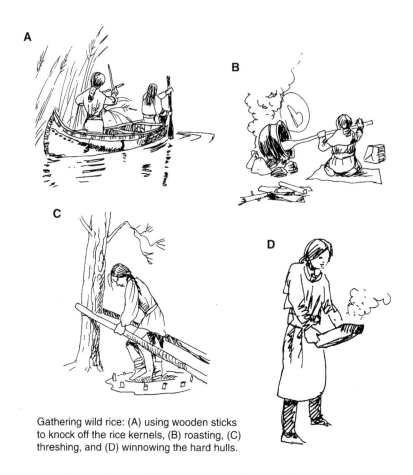

Gathering wild rice: (A) using wooden sticks to knock off the rice kernels, (B) roasting, (C) threshing, and (D) winnowing the hard hulls.

long cold months we all wait for the Wintermaker to fall from the sky and again mark the time for our move to the sugarbush camp.'

"Bineshiikwe," the woman on stage quietly said, "also wrote the following passage:

> *Ojibwe women live on a circle of time, a circle defined by the four seasons of Mother Earth. Somewhere on that circle walks my grandmother, walks me, walk all Ojibwe women.*
> *Throughout the year each Ojibwe woman has many different responsibilities to tribe, clan, and people. But in the end we are all the same, we are women of the land—and that says it all."*

"Thank you for a wonderful reading," Earth Walks said as he came back onto the stage. "Bineshiikwe would have felt very honored."

To loud applause, the woman bowed to the audience and gracefully made her way off the stage.

"Today," Earth Walks boomed, stretching himself over the lectern until he was on his tip-toes, "we can no longer make the Ojibwe woman's claim. We are not women, men, or even people of the land.

"I think," he cried, eyes beaming, "one can reasonably ask if this is a good way to live, or is it a bad way to live?

"We modern people, with our wonderful technology, our incredible science, are able to feed, clothe, and shelter ourselves without the necessity of observing, adapting, or changing as the land changes and the seasons turn. As we stand before a brand new century, it would appear we no longer need to be as spiritually or physically close to planet earth as the Ojibwe were. We have progressed beyond this dependence. So this must surely be a good way to live.

"Yet, like the silver-tongued devil of song," he exclaimed, his voice rolling like thunder, "I could tell you that we have become a rootless people because of this loss. We no longer have a spiritual connection to the planet. And surely this is not a good way to live.

"The Ojibwe and Dakota people I have been speaking about had roots that went deep into planet earth, roots that connected their souls with Mother Earth and all things on Mother Earth. This spiritual closeness to the land made them excellent observers. This is what enabled them not only to make many of their biological and geological discoveries, but also to live well with the land for thousands of years.

"This spiritual closeness also led them to believe that knowledge was inherent in all things. According to Luther Standing Bear, the world was seen as a library and its books were the rocks, trees, grass, streams, and birds and animals that shared, alike with Native Americans, the storms and blessings of Mother Earth.

"My point here is that the technological and scientific separation of humanity from planet earth, our lack of a spiritual connection, has led to a great loss of respect, not only for the planet, but also for each other. This can been seen day in and day out, from the haze and smoke hanging over the rain forests of South America and Indonesia to the clear-cut hills and valleys of North America; from the strip mines of West Virginia and Montana to the acid lakes of Minnesota and Ontario; from ozone holes, melting ice caps, and road rage to beach litter, polluted streams, and racism—and on it goes, a never-ending circle.

"The naked truth is technology and science, born in the Western cauldron of exploitation and domination, are, on their own, soulless, and without a soul there is no spirit, no respect, and therefore no humanity.

"If you believe this is so, then definitely this is not a good way to live. To change this, it is not necessary to stop or curtail technological and scientific achievements, as some would do, for the achievements are nothing less than extraordinary and make our lives so much easier and more interesting. No, the problem is not with the achievements, but with how they have been attained. The change we need to bring about is therefore similar to Dorothy's quest in the Wizard of Oz: to help the tin man, technology, get a soul, and the man of straw, science, find a spirit.

"How do we go about doing this? A hard question, for it has many possible answers. But I can tell you this: one measure of such a change will be when science and technology start to consider the consequences of their actions and achievements— from logging and mining to the building of sport utility vehicles and cloning—on planet earth and the creatures that live upon her. And by consequences, I mean not for today or next year but the next decade; not for us or our children but for their children. When that happens we will know that the spirit of planet earth has secured a resting place in the halls of science and board rooms of technology."

Holding up both hands, Earth Walks finished the lecture by saying, "May your footsteps tread softly on Mother Earth and may your children and their children always have a plentiful, beautiful planet to rest on."

After the lecture, when the auditorium was empty, I went over to Earth Walks and offered him a five-dollar bill.

"What's this for?" he asked with suspicion.

"The sermon," I told him with a crooked smile. "You would have made a great bible thumper, a super television preacher—do you realize how much money you could be raking in?"

He glared at me a moment, shook his head in disgust, and then laughed. "You disagree with something I said?" he asked.

"In substance, no," I replied. "However, I don't think it would play well in the board rooms of Exxon, Boise Cascade, or Ford Motor Company. And it's companies like these that pick up part of the cost of my research. As well, your message was spiritually uplifting to say the least, but to believe the people who drive technological change will trade greed and Yankee dollars for people not yet born is somewhat naïve. But me, overall, I would agree with most of what you said.

"And oh, how you said it, wise sage!" I exclaimed before he could reply. "I'm surprised those people didn't rush right out to kiss the holy earth and tar and feather the first scientist they met."

"Which would have been you," he dryly observed. "But I thank you for your enthusiasm. As for my being naïve, it would be well to remember that sometimes the simplest events or actions have the most profound and unexpected results. It doesn't take a big wrench to stop an incredibly complex machine, and look at what a tiny bug can do to a complex computer."

"Yeah," I replied. "And look at the tracks a sport utility vehicle can make across your forehead."

Laughing, he pocketed my five dollars. "You have a point," he said. "Now that I have a bit of money, let me buy you dinner. We really need to discuss the next lecture. But first you should remember, I did say I didn't have the one and only answer. All I have are my beliefs, and I try to let my life be an example of those beliefs. However, my way may not be your way, since we see planet earth from different perspectives. We each need to find our own earth roots, our own earth soul, and yours can, should, will, differ from mine. But that does not really matter as long as we both begin our search from the same point."

"And that is?" I asked.

"That the interests, the well-being of planet earth and all that live upon her, are put above everything else. If we agree upon that, then we can both stand on the same circle, shoulder to shoulder, both our eyes open, you and me—one with the turning earth."

"You can give my five dollars back," I said.

"Why?" he asked, somewhat startled.

"Because every once in a while you say something really bright and intuitive. So I will buy dinner. But on two conditions."

"And these are?"

"One, blackflies, deerflies, and ticks are excluded from your statement about all that lives upon the planet being placed above everything else."

Laughing loudly, he said, "The second?"

"That our dinner conversation avoids all things spiritual and Native American."

"And geological?"

"You bet," I agreed. "We can discuss the Lake Superior extravaganza tomorrow. Now, do you by any chance have any interest in single malt scotch or California wine?"

"No," he said making a sour face. "I'm much more into fishing, the Impressionists, or soccer."

And so it went for the rest of a rather pleasant evening.

Later that night, while sampling a bit of Cala Ila, I thought about Earth Walks' lecture. I still had a tingling sensation from it that wouldn't go away, a general uneasiness I hadn't felt since I had been given tenure at the university. This was one reason I wanted no discussion of this kind of stuff at the dinner table.

Now I tried to put it into perspective. A lot of my research effort, and much of my early professional career, was concerned with the search for economic concentrations of minerals, particularly those to be found in volcanic rocks. This occupation was, to me, like going on a great treasure hunt, or being a detective and trying to solve a murder mystery with just a handful of clues. One took what geological clues or information could be obtained from the minerals, rocks, and landforms in a particular area, and from this tried to deduce whether there was potential for an ore deposit to occur within that area, and if so, just where it might be hidden. Unlike Nero Wolfe and Hercule Poirot, however, most exploration geologists, including myself, were unsuccessful at solving such a geological mystery. This was in large part because ore deposits, unlike murderers and motives, are rare, and therefore, even with the right geological conditions, are more likely to be absent than present. Even when they were present, even with all the geological clues in place, ore deposits could still be so well hidden beneath the earth's surface that finding one would be like finding a needle in a New Orleans Superdome filled with hay!

I had actually enjoyed a couple of small successes over the years, but during it all I had never questioned whether or not what I searched for, or discovered, really needed to be dug out of the ground at that particular moment in time. I also never worried, or even thought, about how my discovery was to be taken from the earth, nor the consequences of its extraction. The search and discovery, the hunt for the treasure, the solving of the mystery, that was all that interested me.

These thoughts somehow came to rest in the arms of something called geological mapping. Mineral exploration begins and ends with geological field work, and geological mapping is the foundation of all field work. Essentially, geological mapping is the determination of the types of rocks occurring in a specific area, the geological environment they formed in, and their distribution beneath the earth's surface. In mineral exploration there is the added problem of determining whether or not the rocks are potential hosts for ore, and if so, where the ore is to be found.

Now, thanks to Earth Walks, staring me straight in the eye was a thought that had been lurking in my subconscious for years—that geological field mapping was as much an art, an intuition, as it was a science. That intuition, sixth sense, or possibly an innate spiritual connection to planet earth, when combined with good geological training, allowed the right connections to be made—the rocks properly identified, their spatial distribution figured out, the environments of deposition determined, and, in some cases, ore deposits found.

In this regard I wondered whether or not geological observations, when combined with the intuitive part of geological mapping, were not similar to Native Americans' explanations of how they acquired their knowledge of plants and plant use. How this knowledge was based on detailed observations, as well as an innate connection to planet earth, a spiritual tie that somehow directed an individual to the right use, that helped one come to the right solution. Overall, I wondered if planet earth was not a whole lot more mysterious than science could describe.

That thought brought to mind something Albert Einstein once said:

> The most beautiful and most profound emotion we can experience is the sensation of the mystical. It is the sower of all true science. He to whom this emotion is a stranger, who can no longer wonder and stand rapt in awe, is as good as dead.

I believe that. I also believe my sense of wonder and awe concerning planet earth is alive and well. Possibly that was the reason I was sitting here asking myself these crazy questions in the first place.

And just maybe that's what this is all about: to ask questions, and in doing so not to lose that childlike sense of wonder and awe, that bright, wide-eyed view of planet earth that provides the doorway for new experiences, the pathway to unanswered questions and new discoveries.

I hadn't thought about things like this for years. It was pretty scary what Albert Einstein, a philosophical Native America, and single malt Scotch could do to one's mind!

CHAPTER FIVE

THE WOLF'S HEAD

PART ONE

THE FIRE OF MANIDOO AND THE OLD MAN

"There is a vastness about Lake Superior's blue-green waters that make it more like a sea than a lake," I told some two hundred people. Somewhat to my surprise, after Earth Walks' Saturday sermon, the lecture hall was overflowing, with a few people actually standing in the back.

"In fact," I continued, "it's been written that Lake Superior breeds storms and rain and fog just like the sea, making it wild, masterful, and dreaded, as any great sea should be.

"The lake also has character," I said more softly. "The way it gently rubs against the legs of laughing children on a sandy beach, the rise and fall of its sealike swells, deep and regular like the slow breaths of a sleeping giant, and the frantic roar of its storm waves, building higher and higher, as they are driven against its rocky shores by the wild, north wind.

"Then there are its people—people who have lived along its shores for more than six hundred years—people who call it *Ojibwe-Gichi-Gami*, meaning Great Lake of the Ojibwe. To most, though, it is known as Lake Superior, a mistranslation of the name the voyageurs gave it. Paddling their birchbark canoes through its tall blue waves, they called the lake *le lac supereur,* which simply referred to its location above the then better-known lakes of Michigan and Huron.

"Lake Superior, as it turned out, was most appropriate, for it truly is a first-class Gichi-Gami. Superior could fit all of New Hampshire, Vermont, Massachusetts, and Rhode Island and most of Connecticut within its shoreline. The lake measures 160 miles across, 400 miles in length, from Duluth to Sault Ste. Marie, and contains enough water to cover all of North and South America to an icy depth of one foot. Icy, because its average temperature is 40°F.

"It took detailed topographic surveys, and then satellite images, to show us that Lake Superior has the shape of a wolf's head, though it certainly appears to be a Disneylike wolf who finds the city of Duluth full of exotic smells."

"Or it could be the big bad wolf," Earth Walks exclaimed as he came to the podium, "a not too friendly creature just waiting for the right moment to devour the entire city."

Holding up a bow and arrow, he continued, "Long before satellites, airplanes, cameras, or topographic surveys, the North American Indians told stories about the great bow and arrow of Gichi-Gami. From their travels around the lake, and from their geographic and mathematical knowledge, the Indians had

Lake Superior has the shape of a wolf's head.

figured out that the shape of Lake Superior resembled that of a bow and arrow.

"The string of the bow represents the south shore," he said, as he slotted an arrow onto the string and pulled the bow back. "The arrow represents the Keweenaw Peninsula, while the north shore is the bow, which has been drawn back, ready for action. Unfortunately, the stories never do tell us what Lake Superior is supposed to be aiming at."

Putting the bow and arrow aside, Earth Walks continued, "The formation of Lake Superior has been the subject of legends, poems, and the science of geology.

"According to Indian legends Lake Superior was made after a great flood and after the god Nanaboujou had recreated all the big islands and the continents. When the land was whole again, Nanaboujou took his long measuring string in hand."

Earth Walks took a long piece of leather from his pocket, held it up, and smiled broadly. "This is not the original string," he said. "It's one of those scale-model replicas—HO-scale, I think. Anyway, Nanaboujou walked all over the earth, measuring everything with his string.

"He decided on the length of rivers, on the size and depth of the lakes, on the height of the hills and mountains, and on the shape of the land so that everything would be in good proportions.

"When this was done he walked once more over the earth, searching for the perfect place to make his greatest lake. When he found this place—not too far north, nor too far south; forest on one side, prairie on the other—he made his great lake. And he made it in the shape of a bow and arrow, so the people that would live along its shores would always have good hunting."

Pausing, Earth Walks dropped the string and pulled a small leather notebook from his pocket. It looked old and worn. "There are also poetic explanations of its formation," he explained, as he opened the book. "Here is one I kind of like."

And he read:

> *Superior was born and shaped by the fire of Manidoo,*
> *Then finely sculpted by the icy fingers of old man winter,*

And when the warmth of the sun returned to the land
It was filled with the old man's blood;
Which is why its waters are cold and clear as ice,
And why the north wind plays along its rocky shores."

"Finally," I said, taking Earth Walks' place at the lectern, "comes the geological point of view: that Lake Superior was forged by glacial ice, but given birth by the fire of a hundred volcanoes. One billion years ago, in the far reaches of geological time, the land split open like an overcooked sausage. A great crack formed in the earth's crust, a crack that spread from the Lake Superior region, down what is now the St. Croix River Valley, and on through Minnesota and Iowa and clear across Kansas. Geologists refer to this great crack as the midcontinent rift, a place where the western part of North America tried to separate from the eastern part, much as Africa and South America did some 200 million years ago. Hot molten rock—magma, or lava—rose along this crack and poured out to cover the land. Year after year, lava flow after lava flow, thicker and thicker, until a pile of basaltic rock some two to ten miles deep covered the Lake Superior region.

"These lavas, being about as sticky as wet cement, had a great tendency to pile up close to their source, and so they were thickest directly over and adjacent to the great crack. The tremendous weight of this rock pile caused the land to sag downward for hundreds and hundreds of feet. This formed a large, bowl-shaped depression that geologists today call the Lake Superior Basin.

"Today you can walk on these ancient lavas along the north shore of Lake Superior, on the Keweenaw Peninsula of Michigan and on the island of Isle Royale. When you hike over these one-billion-year-old rocks, you will find they look a lot like recent lava flows found on the Hawaiian Islands or in Craters of the Moon National Park.

"Lava flows, like those in Hawaii or the North Shore of Lake Superior, often have round or angular holes in them, which are formed by volcanic gas—mostly water vapor, carbon dioxide, and sulfur—as it slowly rises and escapes from the cooling lava, much as round bubbles of carbon dioxide rise to escape a yeast-water mixture. These holes, called vesicles, vary in size from pinheads to rare ones that are the size of watermelon. Over the eons, as the lava slowly cooled and turned to rock, surface waters, seeping down into the cooling lava, precipitated out a wide variety of minerals that filled the holes to give us such beautiful things as thunder eggs, agates, and geodes.

"In the Lake Superior region, the vesicles were filled with minerals that the native peoples found to be valuable as trade goods, useful in the making of tools and decorations, and important for their magical powers or spiritual significance. These were such minerals as native copper, which you will hear a lot more about in just a short while, native silver from the Thunder Bay area, fine quartz and hematite which form the banded Lake Superior agates, and the so-called greenstone of Isle Royale.

"When volcanic activity finally came to an end, a long period of weathering and erosion began. This erosion created a huge amount of sedimentary material, material that was carried by rivers and streams into the Lake Superior Basin. The erosion of the lavas went on for millions of years, and over this span of time the sediments managed to fill the basin clear to the top, creating a broad, swampy plain.

"Over the next few million years this sedimentary material became hard, solid rock. Several times these rocks were covered by great oceans; during dry times plants and trees spread over them, and they were walked upon by great dinosaurs. Through it all, more than a billion years of earth history, the Lake Superior Basin remained about as flat as a pancake.

"And that's the way it was until the coming of the glaciers about two million years ago. The ice, as it periodically advanced and retreated over the Lake Superior region, found the sedimentary rocks that filled the basin much to its liking, and proceeded to scoop them out as easily as a child scoops seeds out of a pumpkin. The removal of these rocks, coupled with the subsequent glacial plucking and abrasion of the underlying lavas, excavated and deepened the Lake Superior Basin. The glaciers formed an abyss over 1,300 feet deep.

"During the final advance of the Laurentide Ice Sheet, some 20,000 years ago, the basin was completely filled with ice. When the ice began to melt, some 13,000 years ago, the meltwater was trapped between the retreating ice and the southern and southwestern edges of the basin, forming a glacial lake geologists call Duluth. Though covering a much smaller area than the present day lake, Glacial Lake Duluth had water levels more than twice as high as those of modern Lake Superior; this high water drained away through the Brule River, in what is now Wisconsin, to discharge via the St. Croix Valley into the Mississippi River.

"As the Laurentide Ice Sheet continued to melt and shrink, it eventually uncovered a new outlet for Glacial Lake Duluth in the vicinity of Sault Ste. Marie. Because of the tremendous thickness of ice, more than one mile, that had been sitting over the Sault Ste. Marie area, the outlet was lower than the water levels of Glacial Lake Duluth. (Sault Ste. Marie was lower than it is today but due to crustal rebound, this part of the land has risen 180 feet in the past 6,000 to 7,000 years.) The difference in elevation between Glacial Lake Duluth and the Sault Ste. Marie area caused a sudden drop in the water levels of Glacial Lake Duluth, and possibly led to catastrophic floods down the St. Marys River channel. These times of great floods, interestingly

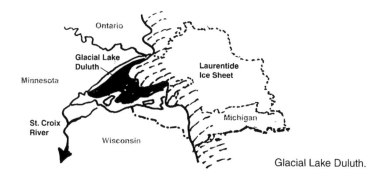

Glacial Lake Duluth.

enough, may be remembered in Ojibwe myths, such as in the story of Nanaboujou and his battle with the giant beaver."

Carrying a large cardboard box, Earth Walks walked to the front of the stage. He opened the box and took out a large skull, which was usually found in the fossil collection of the geology department.

"Giant beaver were real," he said, holding up the skull. "This used to be one of them. They were creatures some six to seven feet long and four feet high at the shoulder. The term 'beaver pond' must have had a totally different meaning way back then!

"In the Ojibwe story a giant beaver has constructed an immense dam blocking up the waters of Ojibwe-Gichi-Gami not too far from *Boweting*, which means Place of the Rapids. It is now called Sault Ste. Marie. The giant beaver and Nanaboujou are archenemies, and Nanaboujou is forever chasing the beaver all across Lake Superior with only one object in mind: to bash its brains out. The two have one great battle after the other until the beaver, tired of fighting and realizing Nanaboujou will never give up, decides to leave Lake Superior. However, the only escape route is to break through its own immense dam. Sensing Nanaboujou's approach, the beaver attacks the dam in a wild frenzy. Just as Nanaboujou creeps close enough to deliver a crushing blow to the beaver's head, the dam gives way. Water rages out of the breach, washing mud, debris, and beaver down

Nanaboujou is forever chasing the giant beaver around Lake Superior.

the St. Marys River and into the southern shores of Georgian Bay. There the mud, debris, and beaver are deposited in the form of thirty thousand islands. Today you can visit and explore some of these islands and see if you can figure out which one was once a giant beaver!

"Now we know giant beaver went extinct about 10,000 years ago during the final melting of the ice sheet. So one can ask if this story is an ancestral memory of mega-fauna, the breaking of a great ice dam, and an ensuing catastrophic flood that represents the end of Glacial Lake Duluth, as well as the end of the age of ice. Or," Earth Walks asked with a shrug of his shoulders, "is it all just a very tall tale?"

"Though it is long gone, Glacial Lake Duluth is not completely forgotten," I told the audience as I watched Earth Walks carefully pack the skull away. "There are old beaches, left high and dry and more than 180 feet above the current lake, there are sticky red clay deposits spread across the region, making gardening in the north land such a joy, and there are steep, rocky promontories, ridges, and shores that were once under cold, clear water.

"The Lake Superior Basin is deep and rocky. It is deep because of volcanic activity, glacial erosion, and the difference in water levels between Glacial Lake Duluth and the Sault Ste. Marie outlet. It is rocky because the hardened lavas were relatively resistant to erosion, so they rise as steep and broken cliffs, barren and grand, high above the restless lake. Present-day Superior only partly fills this basin, and this led to all sorts of problems for people whose main mode of travel and trade was canoe and water.

"For the Ojibwe, the Assiboine, and all the other tribes living in the vast lands to the west and north of Superior, and for the voyageurs and the fur trade, there were only three easily passable routes into and out of this steep, rocky basin. And here I use the words 'easily passable' in comparison to the other possible choices.

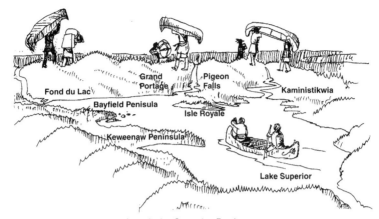

Favored canoe routes out of the Lake Superior Basin.

"The southernmost route, via the St. Louis River, led to the founding there of Indian villages and the fur post at Fond du Lac while the northernmost route, along the Kaministikwia River, saw the building of Fort William. In between the two was the Pigeon River route, which led to Ojibwe villages and the fur post at Grand Portage. European and Indian alike favored the Pigeon River route, for it led, by an almost uninterrupted chain of lakes and rivers, clear to Lake Winnipeg and the plains. Its only major obstruction occurred close to the lake, where the Pigeon River plunges through a basalt gorge in a series of waterfalls and rapids.

"The Ojibwe originally settled at Grand Portage and named it *Gichi Onigaming*, Great Carrying Place. They knew it had been a well-traveled path along the Pigeon River and around the Pigeon Falls for thousands of years—a path that had been extensively used by their prehistoric ancestors and that would continue to be used by those who trapped and traded furs.

"So there it is: Ojibwe Gichi-Gami and the Ojibwe people, Lake Superior, the voyageurs and the fur trade—all woven together and shaped by volcanic fire, glacial ice, and a deep, rocky basin called Superior."

PART TWO

PEOPLE OF THE RED METAL

"Volcanic fire and glacial ice also provided the Lake Superior basin with a very special treasure," I said, holding up a large lump of native copper, "a red metal that is pure, soft, and easily workable.

"Small amounts of native copper occur in many places throughout North America, but it is only in the Lake Superior region, principally on the Keweenaw Peninsula, Isle Royale, and Michipicoten Island, that it is concentrated enough to make the greatest native copper deposits in the world—concentrated and minable because of the fire of Manidoo and the old man's long winter night.

"Volcanic fire—the lava that bubbled and poured out of the midcontinent rift some 1.1 billion years ago to form the Lake Superior Basin—is also host to the deposits of native copper. The red metal occurs in the lavas in three forms:

(1) Round to irregular lumps that have filled in vesicles (gas holes), much as silica and iron did to form the Lake Superior agates. Near the top of some flows, the escaping volcanic gasses came together, like blown soap bubbles, to leave behind large cavities, which were later filled with masses of copper weighing thousands of pounds.

(2) Tabular to sheetlike veins, up to a canoe wide and traceable for hundreds of feet. These represent the filling of cracks or fractures in the flow rocks by the native copper.

(3) Small grains to large, minable masses within layered sedimentary rocks. These water-deposited rocks, which occasionally occur between lava flows, are composed of weathered and eroded lava flow material. This material, as silt- and sand-size grains, was carried by streams and rivers to shallow lakes that temporarily formed on top of the flows.

"The native copper, like the Lake Superior agates, is believed to have precipitated out of hot waters that seeped through the flows after they cooled and hardened. Geologists call this hot water geothermal or hydrothermal water.

"The source of the copper appears to be the thick pile of lavas themselves. It is believed the lava flows contained tiny and widely scattered specks of copper that were leached or dissolved out of the flows by the hot geothermal waters, much as sugar lumps are dissolved into hot tea or coffee. This hot, copper-rich water then rose to the surface along fractures and cracks and through the Swiss-cheeselike holes, or vesicles, left behind by the escaping volcanic gas. Close to the surface, the water cooled and the copper crystallized out, as sugar does from maple syrup, to form vesicle fillings and veins of the shiny red metal.

"Much as did Lake Superior itself, the deposits of native copper had to wait millions of years for the glacial ice to come along and uncover them. The ice did this and much more. It plucked pieces of copper out of the veins and scattered it as boulders and nuggets all over the countryside. When the ice melted away, the copper was left behind as part of the glacial till and as glacial erratics. It was this 'glacial' copper that the Indians first found and used, and it was this copper that led Indian prospectors to the mother lodes contained in the lava flows.

"Probably the most famous piece of glacial copper is an erratic called the Ontonagon Boulder. This three-ton piece of copper sat some sixteen miles inland from where the Ontonagon River, in the Keweenaw Peninsula of Michigan, empties into Lake Superior. The boulder's original home was probably Isle Royale, and its current home is the Smithsonian Institution, to which it was moved in 1843. The boulder bears many wounds in the form of cuts and chisel marks, and from these it is estimated that it once weighed five tons. For hundreds of years before it was moved to the Smithsonian, it was known to the French and English explorers and fur traders, and for even longer to the North American Indians. The Indians believed the shiny red boulder had supernatural powers, or was a gift of the gods. Either way, they left many offerings of food and tobacco for it. The prehistoric copper miners most likely knew of its existence as well, and possibly they also left offerings to ensure good prospecting.

"From 1840 to 1847 the Lake Superior region produced the vast majority of copper mined in the United States; in prehistoric times, from about 3,000 to 7,000 years ago, Native Americans, whom we will call 'people of the red metal' and archaeologists call 'the old copper culture,' mined and supplied copper to Indian bands from Maine to Yellowstone and from the Gulf of Mexico to Quebec. These early Indian miners were a lot like the savvy gold and copper prospectors and hardrock miners of the nineteenth and early twentieth centuries: both left no stone unturned, and in so doing missed very little. It turns out that every modern copper mine on Isle Royale and the Keweenaw Peninsula had already been discovered and worked by these ancient copper miners. It was written in 1873 that the native miners showed 'great intelligence . . . in locating and tracing the veins and in following them up without interruption.'

"When you mine something for 4,000 years you tend to make the countryside look like the prairie dog capital of the world; it has been estimated that there are between fifteen hundred and two thousand mining pits on Isle Royale, and more than three thousand on the Keweenaw Peninsula. These pits, cut into solid rock, varied from less than one foot by one foot to those that were sixty feet deep and thirty feet wide. Some were so large the miners cut drains in the rock to carry off water and keep the pit dry. It is also estimated that the prehistoric miners extracted between 1 and 1.5 million pounds of copper from Isle Royale. How much they took out of the Keweenaw, though certainly more than this, is just not known.

"These original 'hard rock miners' hold a special place in the history of the world," I told the audience. "They were the first workers and users of metal in the Americas and possibly on the entire planet. They owe this happy circumstance partly to the Laurentide Ice Sheet, partly to the warming climate of 7,000 years ago, and partly to their own ingenuity and industriousness.

"It was the glacial ice that scoured away the hundreds of feet of sedimentary rock to expose the veins of native copper, scooping some of it out and scattering it across the landscape like golf balls on a driving range. It was the warming climate that brought a deciduous forest north to the Lake Superior Basin, followed not long after by a people who earned their living by hunting and gathering. As these people moved into the newly established forest, they found themselves moving into an essentially empty environment. Empty in terms of other people, but containing more food on the hoof than a Texas cattle ranch. The woods teemed with elk, deer, barren-land caribou, and lynx; and in the spring and fall the lakes and rivers were home to large numbers of migrating geese and ducks. It is possible that thoughts of succulent goose dinners are what first turned their eyes toward the north country!

"These Native Americans traveled north by boat, following rivers and streams, and as they slowly moved closer to Lake Superior, they could not help but notice the lumps of shiny red metal that lay along stream bottoms and stuck out of the sandy river banks like plums in a plum pudding. Naturally curious, and always on the lookout for anything new or different, they picked some up and brought them home.

"From this small beginning came great discoveries; the red metal could be beaten flat, it was bendable, and it was possible to work it into many different shapes. What started out as a curiosity object, an ornament, or a lucky charm, soon became what Earth Walks calls the magic red metal.

"Dropping chunks of the metal into a hot fire could have been an accident or a planned experiment. Either way, it was a great day in the history of tool making. Heating the red metal and then slowly cooling it made it less brittle and much stronger. And so began the art of annealing—cold-pounding copper, heating it, then cooling the hot metal with water, over and over as it was worked into the desired shape. Using this technique, the people of the red metal found they could make all sorts of weapons and tools, from axes, spear tips, and knives to fishhooks, awls, chisels, adzes, gouges, and crescents.

Cold-pounding of copper led to
the making of many different
kinds of tools and weapons.

"By happy coincidence these monumental discoveries were taking place at the same time many people were beginning to mess about with boats. Bands living along rivers were beginning to explore the waterways past their own territorial boundaries, and in doing so they may have bumped into, and traded with, different bands. As a consequence of this, word of the red metal could have spread along the water trade routes faster than a grass fire on the prairie.

"This would have led to increased demand for the metal, and increased demand meant that more copper had to be found, and so it was that the first prospectors set out to locate the source of the red nuggets. Much like good field geologists of today, these first prospectors were thorough and observant; they recognized that the red metal was sometimes coated with a soft green or blue material. The green was the copper-carbonate mineral malachite, and the blue its sister azurite, both of which form naturally on copper that is exposed to running water or rain, much like red rust (iron oxide) forms on iron exposed to air or road salt.

"The Indian prospectors, following the trail of glacial erratics and nuggets north, exploring the ridges and hills, must have come across green malachite and blue azurite staining and coating the rocks and soils, giving them the look of a painter's palette. Recognizing the connection between their nuggets and these stains, they dug this material away and exposed the red metal underneath. With this discovery there was no looking back, mining of the metal began, and humankind had taken a baby step down the road to industrialization.

"And that's all, folks. These Native Americans, the 'people of the red metal,' never took the second great step. For more than

4,000 years they worked the red metal by cold-hammering and annealing, and they traded it mostly in the raw state; they never learned to smelt or to cast copper, for which you have to melt the metal and pour it into molds.

"These were ingenious and hardworking people, they achieved great things, and they obviously heated the copper in fires, so what happened? Why have they left it for us to speculate about their path not taken? Many reasons have been put forth. If we ignore those that suggest these people were underachievers or technological failures, then one of the more reasonable theories is that by hunting game, catching fish, gathering wild plants, quarrying rhyolite and flint, mining copper, and trading the red metal far and wide, they had no way to support, and no interest in supporting, the specialization and pyrotechnical industry— namely pottery making—necessary to develop metallurgy. It seems that throughout human history pottery making comes after the establishment of some sort of agrarian society, and most smelting of ores comes after pottery making. Pottery making needs kilns and high heat, which turn out to be the exact ingredients for the melting of metals."

"As with most questions of this sort, there could be many other reasons hidden in the mining pits and along the river banks these people called home," Earth Walks offered. "If we take time to look at the question not from the skyscrapers of industrial America, where progress and success are often measured by technological change, but from the valleys and hills of Mother Earth, then the answer might be found in the words of the Sioux chief, Luther Standing Bear:

> The man who sat on the ground in his teepee meditating on life and its meaning, accepting the kinship of all creatures and acknowledging unity with the universe of things was infusing into his being the true essence of civilization. And when native man left off this form of development, his humanization was retarded in growth.

"In other words, these people may have believed they had more important things to attend to, so they had no need or desire to take the next step. They lived in a land with abundant game, fish, and fowl; there were nuts, berries, and wild plants; it was relatively warm, and they had boats and exploring they could go. Could it possibly be that these people were content with what they had and how they lived? Is it possible that here were a people, poised on the brink of a great technological leap, and they simply said no? Possibly life for them was not always wanting more of this or that, or striving to be better than that person, or being upwardly mobile but was, as the Blackfoot Indian Crowfoot said, the flash of a firefly in the night, the breath of a buffalo in the winter time, a little shadow that runs across the grass and loses itself in the sunset, along with the time to realize and enjoy all this. I really hope this was so and that these people were very early practitioners of Luther Standing Bear's words. For not only is it a nice thought, a beautiful philosophy, and a possible lesson for us all, but Mother Earth did allow them to live on the land in their way for over 4,000 years."

At this point, Earth Walks left the stage, and I continued with the rest of the lecture.

"These people lived this way until a dark, rainy day when, for the first time in their collective memory, the mining canoes returned without a full cargo. The surface and near-surface veins and large vesicle fillings were exhausted; there were no more large pieces to be found. From artifacts found in burial sites, we know it was about 3,000 years ago when these people stopped making copper tools and weapons and began to concentrate on small objects, such as beads, hair pipes, rings, awls, pins, and bracelets. Small items require smaller pieces of copper, and much less of it. It was also about this time that they tried to manufacture larger tools by interlocking thin sheets of copper, which would not have been necessary if larger lumps were available. All of this strongly suggests that near-surface vein deposits had been depleted. This depletion of the rich sources of copper must have come as a great blow to these people, for it surely meant the end of trading as they had known it. To make things even worse, it was around this time that the planet's climate turned colder and wetter.

"The combination of these two factors, weather and rock, spelled the end of the line for a culture that had been prominent for almost twice as long as the Roman Empire would be. The end came about 3,000 years ago for the people of the red metal, but not for copper mining, working, and trading.

"To a minor extent, and with small nuggets, copper technology continued into Woodland times. Starting about 2,200 years ago, the Hopewell people of the Ohio Valley mined small nuggets of copper, and with these raised the making of copper ornaments to a great art.

"Indians of late prehistoric and early historic times, the early 'Ojibwe' in the Lake Superior region, also knew about native copper, but they mostly thought of it as a substance endowed with supernatural powers. From about A.D. 1700 on, most of the metal in the possession of the Ojibwe was in the form of nuggets that were worshiped and held in veneration as unusual works of Gichi Manidoo.

"Ojibwe often carried these nuggets around with them in their medicine bags and handed them down from father to son. Larger pieces were considered family treasures and were hidden and kept secret, for these had great magic and the power to aid the owners in hunting and trapping and to bring good health and good fortune.

"So with time it came to pass that the soft red metal, which led the early miners and workers of copper to great technological and geological discoveries and gave them prestige and power, ended up becoming an emblem of worship for the Ojibwe. An emblem that in the mid-1800s would start a copper rush, bring miners by the trainload to the Lake Superior region, and eventually cost the Ojibwe millions of acres of land.

"We will now have about a fifteen-minute break. You'll find coffee, soft drinks, and tea, along with sweet rolls, in the alcove immediately across from this lecture hall. After you have been properly and adequately refreshed, I would like everyone to gather at the rear entrance to this building. If you're not sure how to get there, I will be glad to show you at the appointed time.

"Once outside we will board buses, which will take us over to the outdoor theater, about ten minutes from here. There, on this beautiful afternoon, we will have the conclusion of the Lake Superior copper story and today's program."

PART THREE

MINONG

Though I knew what was waiting for us at the amphitheater, I had never actually seen the set, and I must say, it was certainly impressive. Three wooden panels, each about ten feet high and six to eight feet wide, had been set side-by-side to form a large semicircle on the elevated stage. These were painted to represent the inside of a prehistoric copper mining pit. Wooden ledges, painted to appear like rock ledges, projected from one of the panels. These were approximately two feet apart and rose in steplike fashion, right to the panel's top.

Placed within the semicircle of panels, so it appeared to be a projection of one of them, was what looked like a large outcrop of basaltic rock. This had a reddish-colored vein, about two feet wide, running right down its center. Both the vein and the basalt had been constructed from several separate pieces of sheet metal that had been bolted together. The vein was made out of two such pieces that were hinged where they joined. At the lower end of the vein, where it came in contact with the stage, was a small lever. When this was stepped on, it caused the lower half of the vein to open upward.

The vein was hollow on the inside. Hidden within it was a large copper boulder, also constructed of sheet metal. The boulder sat on an inclined piece of wood and was held in place by a metal band. When the lever was stepped on, the hinged part of the vein swung upward, releasing the band, and the boulder rolled onto the stage.

This entire structure, basaltic rock and vein, sat on a large, flat piece of sheet metal that was painted the same color as the basalt outcrop.

Small pieces of wood and birch bark had been placed along the basalt outcrop, right where it joined the red vein. A small pile of wood was stacked against one of the panels, and several birchbark pots sat next to the wood. Leaning against one of the other panels were two sledgehammers. Only these were not the modern iron or steel ones, but were made of oblong basalt boulders attached by leather strips to long, wooden handles. Next to the stone sledgehammers sat a long-handled wooden scraper, much like the ones used to rake chips off a roulette table.

Once everyone was settled into their seats, Earth Walks came onto the stage. He was again dressed in what he considered to be the appropriate clothes for the time and place. His long shirt and tight pants were made of deerskin, and on his feet he wore leather boots that came half way to his knees. These were tied around his leg with colored strips of fur. Around his neck was a copper necklace, which had a large, elongate object hanging from it.

Facing the audience, he said, "It is here, in the open, that we will continue and conclude the story of native copper and Native American copper miners. I have taken the liberty, or possibly I should say the earth spirit has touched me, to write the following story. This will be dramatized by me and two talented youngsters.

"What we will portray is the way geologists and mining engineers believe Native Americans mined native copper. The rest of our story is open to interpretation, for our knowledge about these people is limited—their footprints have been largely erased by the dust of time. However, there is nothing I will say or do that is contrary to any knowledge we have about them."

With those words Earth Walks turned and started to climb up the ledges. At the same moment his two assistants, boys I guessed to be twelve or thirteen years old, came onto the stage and started placing more wood onto the metal basalt.

Reaching a ledge about halfway up the panel, Earth Walks crouched down and nodded in my direction.

With this signal I walked onto the stage and took my place to the right of the far panel. Clearing my throat, I took a small piece of paper from my shirt pocket. I turned on my microphone and began.

Scene. *Earth Walks, (as Miskwaabibkoke, or* He Who Mines Copper), *crouches on a narrow ledge studying the shining red vein. The vein is as wide as a man, and it runs at a gentle angle across the pit floor and up the far wall. The way the red metal disappears into the blackness of the pit bottom, and the way it is enclosed by the dark rock, reminds him of the long, red trails, some lasting only a heartbeat, that he sometimes sees streaking across the night sky.*

NARRATOR: In 1664 a French trader named Pierre Bouchard wrote:

> In Lake Superior there is a large island about fifty leagues around in which there is a fine mine of red copper; there are also found several places large pieces of this metal in a pure state. They—traders—have told me that they saw an ingot of pure copper which, according to their estimation, weighs 800 lbs. which lay along the shore.

Native Americans called this island Minong, and more than 6,000 years ago they mined copper there. The following dramatization takes place on Isle Royale, located in the northwestern part of Lake Superior. The time is summer, some 3,400 years ago.

MISKWAABIBKOKE (EARTH WALKS): I've been working this vein for the entire season. I feel as if I know it as well as I do my own body. Tcianung, the other cutter, and I, along with fourteen workers, have sunk a series of inclined pits that intersect this vein at greater and greater distances below the earth's surface. This current pit is the deepest yet, over five men deep, and it will be the last.

With Manidoo's blessing, it might also be the best. The way the vein slopes, its width—with the proper fire and right cutting, it could yield the largest piece of the season.

All the preparations are made. The wood has been placed along the length of the vein, at the exact spot where it joins the hard, dark rock. Extra wood, to be added to the fire if needed, is stacked against the pit wall, and birchbark baskets, filled with cold water carried from the lake, sit directly below me.

All that is left is to honor the spirit of the metal. What I wear around my neck is a lump of the red ore, streaked with silver that has been hammered and rolled into the exact shape of this island. It was presented to me by the great cutter, Asiniokumig, the day I joined the ranks of the master cutters. Faithfully I have worn this for every cut I've made over the past fifteen seasons.

(Lifting the metal piece to his lips, Earth Walks gently kisses it, then prays loudly for success. When he finishes, he gives the order to light the fire. Within minutes the dry wood piled across the rocks is burning high and bright, and the rising smoke swirls around Earth Walks, obscuring him from view. When the wood is reduced to hot charcoal, Earth Walks climbs down to the stage. Walking to the far panel, he picks up the long-handled scraper and begins to brush the coals away. In no time at all the rock is completely cleared off.)

All that is left is to honor the spirit of the metal.

(To the two assistants.) As you can see, we have been blessed this day. One burn, when it usually takes three or four to make the metal glow like a cranberry. And look, just look, at the black burn marks across the dark rock!

But this shouldn't hold your interest for long. What you look for are cracks and breaks that radiate from the edge of the vein outward, into the rock, much like the cracks a walking person makes in the thin spring ice. See here, like these and these and these. *(Excitedly, pointing at several different spots along the dark rock.)*

NARRATOR: Satisfied with what he sees, Miskwaabibkoke covers his face with a leather mask, and slides cut pieces of elk hide over his fingers. *(Earth Walks performs the actions narrated.)* Selecting just the right place, not too far up the vein, nor too far down, Miskwaabibkoke takes a water basket and empties it onto the hot rock.

> *(When Earth Walks does this a cloud of steam rises instantly around him. As the steam washed over him, one of his assistants hands him another basket, which he pours just below the first. And he continues in this manner, pouring up and down the vein, until the water is gone.*

> *Stepping back, he removes the mask. His face is dripping with sweat, and he gasps for breath in the hot, moist air. Bending down, he appears to study the vein as a hunter studies the track of a deer. Shaking his long, braided hair, Earth Walks turns and faces his two assistants.)*

MISKWAABIBKOKE (EARTH WALKS): The fire of Manidoo has truly blessed us. The cracks are deep and angled. If Mother Earth pleases, and I am steady and skillful in my cuts, this could be the biggest piece of the season.

> *(He turns back to the vein and appears to concentrate hard on what he is going to do next. Opening a soft deerskin bag, he withdraws stone and metal wedges and chisels, and a beautiful hammer. The hammer is a finely worked piece of native copper*

Miskwaabibkoke fractures the rock using fire and cold water.

about the size of a closed fist. It has been hammered flat and polished on one end, while on the other it was slimmed and honed to a sharp point. With the tools in hand, Earth Walks sets to work widening and connecting the cracks.)

NARATOR: Slowly Miskwaabibkoke works down one side of the vein, and up the other. He joins the cracks together in a jagged, toothlike pattern, much in the manner his wife sews together pieces of deerskin. This connecting of the rock is his life and soul; it is his art, and he has become the best there is. He alone is able to cut and widen and connect so it takes only a few strokes of the long stones to free the red metal.

Totally absorbed, Miskwaabibkoke works without pause until the last crack is wedged and chiseled to the right width and connected to the other cracks.

(With a satisfied grunt Earth Walks sits back on his heels and wipes sweat from his eyes. Looking upward, he is surprised.)

MISKWAABIBKOKE (EARTH WALKS): The long shadows of the sun run down the pit wall. It has taken me longer than I planned, but then the nature of the rock has allowed a much longer cut to be made than I had thought possible.

(Stretching his back, Earth Walks cleans and wraps his tools before returning them to the bag. Standing, he shakes himself like a wet dog, then slowly moves to the back of the pit. Slumping against one of the panels, he takes a long drink of water. He pulls out a leather tobacco bag and fills a clay pipe while watching his two assistants set about their job.

His tobacco burning, Earth Walks begins to count the number of blows the two pretend to deliver to the dark rock with the long-handled stones. Slowly they work their way down the imaginary system of cracks. When Earth Walks reaches fifteen, it happens. Suddenly, with no warning, a large piece of the vein flies upward and a big chunk of native copper rolls out onto the stage. One of the helpers screams, and both leapt out of the way faster than rabbits chased by a fox. The copper chunk rolls across the stage, finally coming to rest some two feet from where Earth Walks sits.

Breathing heavily, the two workers stare at the huge slab of copper as if it were the most wondrous sight they had ever seen. They turn toward Earth Walks, raise their arms, and bow. Throwing down their long-handled stones, they begin to dance around the copper slab, shouting with joy, and calling long and loud like large wild cats.

Lifting his head, Earth Walks blew the smoke from his mouth and screamed long and deep at the dark blue sky.

MISKWAABIBKOKE (EARTH WALKS): The largest of the season! Look at the knobs and points and edges it has, it will be enough to fill the last two canoes. We can go home! *(He jumps to his feet and hurries over to touch and gently rub the slab's ragged-looking surface.)* Two arms long and half a canoe wide. A masterful cut thanks to Manidoo. This one will be talked about for ages. This is the biggest piece I have seen in the last eight seasons!

NARRATOR: Now comes the backbreaking job of getting as much metal off the piece as possible. These smaller pieces will be carried from the pit, down the ridge, to the lakeshore, where other workers will beat them flat for transport home. Miskwaabibkoke's part in the operation is over.

(Slowly, as if his legs and arms were made of cement, Earth Walks climbs onto the first ledge. As he reaches the third ledge, his two assistants are joined by several other boys. In no time they have slid the metal basalt knob off the stage and have completely reversed the three panels.

Coming around the edge of an end panel, Earth Walks stands in front of what is now a breathtaking panorama. He appears to be on a high, rocky ridge that is covered by low shrubs and stunted pine trees. The ridge descends toward the lake in a series of gentle rolls before ending at a wide, rocky beach. Beyond the beach is blue water, flecked with small white caps. In the far distance there is a smudge of solid land.)

MISKWAABIBKOKE (EARTH WALKS): *(To audience.)* After the stale, smoky air of the pit, the freshness of the open ridge tastes good. I am tired, bone tired, and yet, for all my aches and pains, it is good to know I am still the best.

Manidoo, what a beautiful cut, and what a way to bring all this to an end, not only for this season, but forever. I will make no more trips to this place far from home.

I am old, the oldest here. Shingwauk and Nagaunub had not been born when I made my first trip to the island as a worker. But it is more than age. It is also the lake and sky, every year colder and wetter and wilder. I used to love and welcome the rain and the great waves crashing in off the lake. But it was the wind I especially liked, the way it screamed like an eagle as it flung mist across the island. Now, these make my bones ache.

However, I must admit the main reason this will be my last season is the ore. The red metal is nothing like it was twenty seasons ago, and even then the cutters said it was nothing like that which the ancients mined. And these words were true. I know, for I have climbed into the pits of the ancients. I have seen the great pieces of metal, chunks three times my height and as thick as a canoe. It would take ten of the pieces I have just cut to equal one of the ancients.

The ore we mine now is thin and miserable, and I doubt another vein like the one we discovered last season, mined all this season, the vein that has just yielded its largest piece, will ever be found again.

Along the ridge, just there, I can see the other two working pits. I should go along and tell Tcianung that we have enough metal. There is no longer a need to work the thin, miserable ore in those pits. But all in good time. Standing here in the open air feels good, and a short wait would make no difference. (Peering down the ridge.) I can just make out the four whose job it is to beat the pieces of metal flat. Whenever I watch them I wonder whose idea it was to tie the grooved ends of the large, heavy stones to tree saplings. Then, to bend the saplings back as far as they can go before releasing them, sending them whipping forward to come down hard and sure on the red lumps of metal. In this way it takes little time to make the metal as flat as a canoe paddle.

Though I can see they work hard, I can't hear the sounds their great stones make. They are drowned out, like a wet fire, by the striking noises of the two in the pit below me. These men use handheld stones, worn smooth and shaped like an eagle's egg by the water of the great lake. There is no need to attach handles to stones such as these; short hand strokes work just as well on the soft metal.

Further along the shore, just below our camp, are the caches— shallow pits we have dug into the rocky soil and lined with hides. We place the flattened pieces of metal inside these, and when a cache is full, it means a canoe has a full load for the journey home.

Home, *(sighing deeply)* it makes me think of the spirit canoe, and the sadness it will carry. The four men, two miners and two hunters, all my friends and work mates, who have died this season. Two were crushed in the pits by a large slab of dark rock, and two were lost to hunting mishaps. We cleaned their bodies, painted them with the red and silver powder, and then offered them to the Manidoos in the fire ceremony. After the burning their bones were carefully collected and placed in sacred bundles so they could be taken home in honor and respect.

But it will be a sad time for those who wait for our return. I can already see the anxious faces of the families as the canoes round the final bend in the river and head for shore. Before we even touch land, the wails and anguished cries will start.

Bah. Enough of these bad thoughts. Fool—think of good things like the last cut; the amount of metal we take will be enough to fill two canoes, possibly more.

(Breathing deeply, he picks up a small lump of native copper from the stage floor.) It is hard and heavy in my hand, cold and without shape. Even now, at my age, I am still amazed at how easily it is flattened and rolled and worked, and how the craftsmen make such wondrous things out of it. The axes, wedges, knives, spears, fishhooks, rings, and so much more by hammering the metal, heating it, and hammering again, over and over until they have the shape desired. Then the metal can be ground smooth and sharpened, or polished to a bright red color. For me this is indeed magic.

And it is this magic metal that has, for ages and ages, brought my people respect and made them some of the best river traders. All the bands want the metal, and most want it in the raw, unworked state. We have to do nothing more than deliver it to the many villages set up along the rivers and streams. We also take it south, trading down to where the two great rivers meet, and on rare occasions we have gone all the way to where the large salt water rises up, onto the land. We also take it north and east, across the great lake, and some years we even go west, to where the prairie grasses grow, and the great brown beasts cause the ground to shake. Because of our trade over long distances, some have come to call us the Ottawameeg, meaning special people who trade across boundaries, and do so through all of time.

The only real work in all this trade is done by the cutters and rock workers. Without us there would be no trade—no shells from the salt water, no brown robes from the prairies, no silver paints or clay pipes or pots from the great river, no tobacco or the hundred other things we have because of the shining metal. My work, all the cutters' work, is the lifeblood of my people. Because of this we cutters have earned a special place in the band.

NARRATOR: Native Americans mined and traded a red metal called copper for more than 4,000 years. Four thousand years of living on and with the land, and all planet earth has left for us to remember them by are pits, charcoal, hammer-stones, copper, flint, and wooden tools, and a master cutter called Miskwaabibkoke and his hard-rock miners.

The applause was long, loud, and enthusiastic. Earth Walks stood in the middle of the stage and lapped it up like a cat in a cream pitcher. After several seconds he gracefully bowed, then held up both hands to quiet the audience.

When you could once again hear the wind in the trees and the singing of the birds, Earth Walks said, "As usual there is a touch more to this than an historic or prehistoric dramatization. There is a small point. What I hope you will take away with you from today's lecture and this performance is more respect for our ancient ancestors. That you will no longer think of them as many of us tend to do—barbarians, ignorant savages barely able to exist on the land, let alone live well with it.

"This presentation, I hope, demonstrates that these people were not much different from us. They too worked hard to improve their lives, and they did so by using technology. Just look at the geological, engineering, trading, and exploration discoveries they made—they may have been the planet's first entrepreneurs, geologists, engineers, and traders. And we can only imagine what other things they did and knew about that are forever lost in the obscuring mists of time.

"But, as we tried to point out in this morning's lecture, they may have differed from us in one respect. They may, and here I stress the word 'may,' for it may be just wishful thinking, have realized when to say no—that enough is enough. Possibly they had all the material things they wanted, everything they could use. They may have lived what we call the good life of the thirty-fifth century—B.C. that is.

"In comparison, look at us and, as an example, the computer industry in the twenty-first century—A.D. that is. We have a vast computer industry that day in and day out tries to cram technological advances down our throats—and for what? To be better, faster, ahead of the game—a few nanoseconds here, a meg of RAM there, a few more bytes everywhere. It's nonsense, and we swallow it memory chip, video card, and hard drive.

"Technological achievements are great things, but we really need to wake up and realize they are not the end-all and be-all. There is something out there called life, and many of us need to megahertz down, turn on our own memories and imaginations and damn well enjoy it."

CHAPTER SIX

A Long Winter Night

PART ONE

TAMBORA AND THE LITTLE ICE AGE

"As you know, we will conclude this lecture series with field trips tomorrow and Saturday," I told more people than I was certain we could accommodate. This large number of people had made scheduling field trips a nightmare. For the first time I truly wished for a smaller audience.

Out loud I said, "Due to scheduling, hours of operation, and access, we will first visit an Ojibwe winter camp and a North West fur trading post. These have been reconstructed to look as they would have in the late seventeen or early eighteen hundreds. Then, on Saturday, we will go to an Ojibwe maple sugaring camp, as well as look at glacial landforms and volcanic lava flows. You will also have the opportunity to hunt for agates and explore for native copper.

"With the field trips in mind, particularly those to the various historic sites, what we thought we would do today is take a look at the climatic conditions during the time of the fur trade and examine a couple of different geological factors that greatly influenced these conditions. We will also see how Native

Americans in this part of North America coped, lived with, and survived the climatic extremes we will be discussing.

"Speaking of climatic extremes, in Norse folk legends, the chilly breath of the frost giants blew over the land, and a giant wolf swallowed the sun. In North America, the old man of the Ojibwe and the north wind roamed freely across the land, setting loose the time of the long winter night. It was about A.D. 1300, and the Little Ice Age had begun; it was to last 550 years, to about 1850. This time of cold was to have a dramatic effect on people who traveled by water and lived off the land and on businesses, like the fur trade, that depended heavily on climatic conditions for their success or failure.

"During the Little Ice Age snowlines—the elevation at which snow remains on the ground all year round—moved 600 to 1,400 feet closer to sea level, and in Scotland the elevation at which one could practice agriculture was lowered 650 feet. This meant disaster for the people of the Scottish Highlands, and led to the establishment of a Scottish colony in the Red River Valley in 1812. All across Europe crops frequently failed, famine was widespread, and on several occasions the Baltic Sea froze solid and people could skate, slide, and ice-fish from Germany and Poland to Sweden. In North America, during the winter months, you could walk from Staten Island to Brooklyn, and in the Lake Superior region it was said you could walk from Grand Portage to Isle Royale.

"The coldest part of the Little Ice Age was the years 1645 to 1715 and 1810 to 1820. In North America the Hudson's Bay Company kept journals of occurrences at each of their trading posts or houses from 1786 to 1911. Records from 1815 to 1819 show that spring was late each year, summers were cool and short, autumn came early, as did cold weather and ice, and both hung around until late spring to make certain the cycle repeated itself.

"As it turned out, no one really knows if its success was cause and effect, shrewd business sense, Gaia, or plain luck. The Hudson's Bay Company received its charter in 1670, during the coldest part of the Little Ice Age, and a mere ten years before one of the coldest decades in the last several hundred years. Imagine what would have happened if the Hudson's Bay Company had gone into the fur business during a time when the climate was warmer, such as it was during the climatic optimum from A.D. 950 to A.D. 1250. The fur trade would have been a flop. Can you imagine anyone buying a beaver-skin bikini, let alone suntan lotion made of beaver fat!

"Much like Bill Gates and Microsoft, however, Pierre Radisson, Sieur des Groseilliers, and the Hudson's Bay Company were in the right business, at the right place, during the right time; the company prospered, and demand for furs did nothing but increase.

"Unlike many other businesses, the fur trade was almost totally dependent on climate for its health, well-being, and success. Climate dictated the number and quality of furs, conditions, and degree of success for the Indian trappers and their families, and the ease of transporting furs to trading posts. Overall the colder, snowier, wetter, and icier it was, the more difficult to trap, the harder to transport what you did trap, and the longer to get where you wanted to go, all of which meant less profit for the traders.

"When the dependence on climate was combined with the intense (and sometimes bloody) competition for furs between the Hudson's Bay Company and the new outfit in town, the North West Company, it spelled big trouble for the Indians, their land, and its resources. Because of this competition, both companies began to trade for furs at their source, where the animals were trapped, rather than have the Indians bring them to a central location, as had been Hudson's Bay tradition for 110 years. This change in the way of doing business led to rapid expansion into the North American interior by both companies. This placed great pressure on the land and its resources, principally through the continuous depletion of large game like moose and caribou.

"Combine the cold climate of the Little Ice Age with the greatest volcanic eruption in 10,000 years, and the stage is set for a cultural and human tragedy in fur country.

"In the spring of 1815 the big news in Europe was Napoleon, who had escaped from his island prison and was headed toward his Waterloo. At about the same time a volcano called Tambora caused its own Waterloo by blowing up as no volcano had in recorded history.

"Located on the island of Sumbawa, in what is now Indonesia, this 13,000-foot-high mountain vanished, leaving in its place a ragged hole, which geologists call a caldera, some three and one-half miles wide and over 4,000 feet deep. The top of this caldera stood at an elevation of 8,500 feet; the other 4,500 feet of mountain had vanished into thin air. Zillions of tiny bits of lava, along with millions of tons of sulfur, chlorine, and fluorine, had been blasted clear into the stratosphere. Quickly encircling the globe, these lava particles and gas droplets formed a curtain, which reflected enough sunlight back into space to make an already cold planet colder.

"In the wet fall and cold, harsh winter of 1815–1816 famine paid a visit to northern Minnesota, Ontario, and Manitoba. This time of severe hunger affected both native peoples and Hudson's Bay and North West Company employees and their families. There was another famine in the winter of 1816–1817, but not as severe as the previous year's.

"The time of the great famine began in May, some two weeks after the eruption of Mount Tambora. The month of May was very cold, 'more the appearance of March than May,' one person wrote on May 8, 1815. The main body of migratory birds, mostly geese, arrived in mid-May, one month late. June saw dramatic weather changes, from rain and cold to very hot and dry; it was so hot for two weeks near the end of June that the woods burned. In mid-August it turned rainy and cold, and it stayed that way until the snow came. Fall fishing was a disaster, as was ricing, because of high water levels.

"As James Slatter, the master of Osnaburgh House, wrote on September 23, 1815:

> *The water being so remarkably high at this place the Indians is not made any rice worth while . . . so that I am much afraid*

of starving in the winter as there is no fish to be got here when the water is high in the fall.

"It began to snow on October 22, and, though no one guessed it, winter had come to stay. The failure of the fall ricing and fishing harvests, when combined with the depletion of moose and caribou due to overhunting and the seven-year crash of the snowshoe hare population, meant starvation and famine were about to rear their ugly heads.

"Hudson's Bay Company journals note the first cases of starvation of Ojibwe on December 1, 1815, and by January 1816 it was reported to be general among the natives. At Osnaburgh House, on Lake St. Joseph, Hudson's Bay Company records indicate 30 to 40 people out of an Ojibwe population of 180 to 200 starved to death during December and January.

"Those who survived that horrendous winter found the following summer cold and dry. In fact, all across Europe and North America 1816 has come to be called 'the year without a summer.' Summer temperatures averaged 3–5°F below the two-hundred-year average, and June was 5–6°F below normal. The crop of Indian corn, or sweet corn, which was the staple crop of

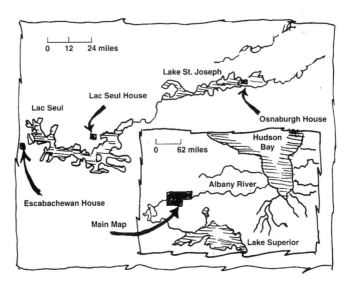

Location of Osnaburgh House, a Hudson's Bay Company trading post.

rural nineteenth-century New England, was a disaster. Frost occurred every month, and a blizzard on June 6 dumped six or more inches of snow across New England. It also snowed on July 9, August 21, and August 30. On July 4 the high temperature in Georgia was 48°F.

"In fur country the summer was extremely dry and very cold. For these reasons water levels fell and remained low. It was written: 'All small creeks that flowed with plentiful streams all summer have entirely dried up.' What was previously shipped by boat, namely furs and supplies, now had to go by cart, and so it was that the famous Red River Carts got their start.

"The climatic conditions during these years seriously affected wildlife, including fur-bearing populations, and the harvest of wild berries and other fruit, rice, and plants. It hindered movement for trade purposes and for hunting, and it ruined the Indian fishing industry. Overall, there was a decrease in game populations, a decrease in fur-bearing animals, and a decrease in Indian populations. The decrease in the Native American population was given a helping hand by epidemics of smallpox, measles, and whooping cough, diseases for which Indians had no natural immunities.

"And that is pretty much the way it was from 1815 to 1820 in fur country, with geology, culture, and climate dictating where, how, and even if the native peoples would live. Gossamer threads, all woven together, held and gently tugged by planet earth."

PART TWO

SHOOTING THE WINTERMAKER

"Those were harsh times indeed!" cried Earth Walks. "We have previously seen how the Ojibwe were a people of the seasons. In keeping with the theme of today's lecture, I would like to show you what life was like during their harshest season, winter. To do this we will journey to an Ojibwe winter camp and spend the long, cold months with an Ojibwe family.

"The story you are about to hear, called 'Shooting the Wintermaker,' takes place during the Little Ice Age, not long after the eruption of Mount Tambora."

Scene. *A large movie screen along the back wall. A pale white spotlight lights one corner of the stage. An elderly woman, wearing winter dress and a string of brightly colored beads around her neck, strides into the white light and faces the audience.*

WOMAN: I am an old woman now. I can hardly see to do my sewing, much less remember where I last set it down. What I do remember is my girlhood, and those years are as clear to me as the stars on a cold winter's night. Like most of us Indians I tell nothing of terrible things; enough to say that before the bad time, things were memorable and lovely, and I especially recall the year of the long cold, the year of my coming of age.

My family agreed to hunt and trap that winter with mother's sister's family, in which there were two girls slightly younger than I. I was excited because I knew that after the winter in the bush, we would become good friends, almost sisters. Living with them was an arrogant boy named Gayaazhk, who teased us constantly and bragged incessantly. He was not of our blood; he was the kind of boy who should have been left on an island by his parents, and maybe that is how he came to my cousins and joined our camp. Father thought he would be useful and help the men get the beaver.

The tamaracks were turning the color of fresh buckskin, and ice had already begun to form along the shore of Gichi Gami, when we set off. What a sight we were!

(The screen comes alive with the many different shades of green, yellow, brown, and red that color the Lake Superior shoreline in the fall. Slowly the camera zooms in on four canoes as they make their way along the rocky shore.)

See! Father in his red blanket coat, and mother's beads sparkling against her black stroud winter dress. She paddles in the stern of a winterbark canoe with scraped floral designs along its sides, and baby Anung faces her in her cradleboard. Gayaazhk and my two cousins are in another canoe, and I, along with my six-year-old niece, paddle with old Adaa-weikwe, the trading woman. In all we are four canoes, and each is laden with rawhide sacks of wild rice, pack baskets, bundles, blankets, cedar mats, rolls of bark, and much more. In the prow of two of the canoes lean the men's long guns, each with a slender stock, and sea monsters carved near the triggers.

> (*The video moves from one birchbark canoe to the next as the occupants are described. From this point on, the video brings to life the story told by the Ojibwe woman.*)

Oh, how I remember . . . our dogs ran along the shore, and we rendezvoused at each portage. At our fourth portage we saw the warning signs painted in red on the grey rocks. There was a picture of the great panther, and below it the warning that when he rose in the eastern sky, the snows would melt and this place would be flooded. Mother told me not to worry, we would be well to the south before the spring waters covered the land.

But our journey now took us northward until we came to the Little Dog River. Turning eastward, we paddled on its slow-moving waters, avoiding the half submerged logs that

In all we are four canoes, and each is laden with rawhide sacks of wild rice, pack baskets, bundles, blankets, cedar mats, rolls of bark, and much more.

floated downstream, while watching the long formations of geese as they gracefully fled the northern sky.

After paddling most of the day, we came to a small lake, and on its far shore, in the midst of a stand of great black spruce, we saw father's blaze, and lob tree marking it. We had reached our winter camp.

I helped the women set up the winter lodges, filling pits with rocks that would be our hearths and help retain the heat. Next we prepared the walls of our lodge, and these we made out of two layers of birchbark with moss stuffed between. Finally came the floor, big and roomy and covered with pine boughs. These we would change each day to keep the lodge smelling fresh and sweet. The moose were in rut, and soon there was fresh meat in the caches built high up in the trees.

Late in what had been a cold, rainy fall it started to snow. The snow was early, but we had nothing to worry about. We made snowshoes and toboggans from the forest. The snare line trails, which we checked each day, were tramped down firm and hard, and father and the other men began to bring in mink and muskrats. I could skin them almost as fast as mother, fleshing as I went, pulling their skins over stretching boards to dry. The men scattered the carcasses along the packed-down trails to feed fishers and weasels, which they also trapped. How proud mother was of me—I could see it in her grey eyes—and I felt myself growing into a sturdy copy of her.

I remember the night the Wintermaker rose in the eastern sky, lifting clear of the horizon just after sunset. Although it had been snowing for days, this marked the real start of winter. The lake had frozen deep and fast, and as I watched the Wintermaker crawl across the night sky, the great summer town of Odanah, with all the visiting, and ceremonies, and games, seemed a dim and far-ago memory. It was easy to imagine we were now the only people in the whole world, alone, small and fragile, under the cold, watchful gaze of the

Wintermaker. The one good thing about the coming of this great star figure was that he brought us deep snow to bog down the moose, making hunting that much easier.

The Wintermaker also brought us hares, not as many as usual, but enough to make warm blankets, coat liners, gloves, and socks. We would cut the thin rabbit skins, fur still attached, into strips, braid them, then weave the braids into what we wanted. We spent the long nights sewing, repairing equipment, and listening to old Adaaweikwe's stories of Nanaboujou. I especially liked the ones about how he defeated great giants, like moose and beaver, who once walked the earth. Life was so good and so full, I forgot to practice my fasting. Perhaps it was this that brought on all that followed?

Or was it Gayaazhk, the braggart, the one dark cloud of my life. Thankfully he slept in the other lodge with his adopted family, and he was gone all day with the men, or I would have gone crazy. His teasing and bragging went unabated, and it continued all through the first part of that winter.

One day my uncle reported a new kind of owl sitting in a tree near our camp, and the men became worried. Perhaps it was an evil shaman casting a spell on us, or possibly an omen foretelling death in our camp. Whatever the owl's purpose, it took the joy and spontaneity out of everyone, even the children spoke in hushed tones.

I remember the day the men came home with no game. And it was the same, day after day. We still had food stores, but after several days we knew this foreboded hard times to come. We used up our dry berries and nuts, rice began to run low, and soon it had to be rationed along with the little meat we had left. Before long I felt what I hadn't felt for a long time: hunger pangs.

Each evening mother would glance at father, a worried look on her face, but she never spoke a word about the lack of food. Father would just sit and stare into the fire, possibly thinking about the long hard day on the trail, and not a scrap of meat to show for it.

The small children cried more often now. They were very hungry, and something had to be done.

The men built a lodge of spruce boughs some distance off. They sang and drummed for days, and we women were forbidden to go there. Even old Adaaweikwe, who was in charge of the camp, sat in silence.

Late on the fourth day, while mother and I were out gathering spruce boughs for our lodge floor, my father approached us.

It is over," he said, in a firm voice. "The moose will come, many more than we need."

Giving me a sharp look, meaning for me to stay right where I was, he took my mother by the arm and led her away. My sharp ears, however, picked up every word he spoke. "It was Gayaazhk," my father said, shaking his head. "He boasted and joked on the hunt early last month. He talked and talked to the trees and squirrels about how many moose he would kill that day. The moose ogimaa heard him and was greatly offended. Gayaazhk has confessed his sin."

"What will happen?" mother asked, a worried look on her thin face.

"It has already happened," father answered. "He has given his best dog and his knife at the offering place. He has taken a vow to atone for this great sin, and he has gone off to do so. Be at peace, all will be well."

Three mornings later we heard a shot far across the lake. Mother, with a great sigh of relief and the first smile I had seen on her face in weeks, grabbed her toboggan and started out. There was a feast of moose meat that night, and for the first time in days the children slept quietly.

Deep into the winter the beaver season began. Father built log traps and snares, and soon our camp was encircled by birch saplings with beaver hides stretched on them to dry in the cold air. The beaver carcasses were roasted on the open fire, and the mild sweet flesh tasted wonderful after our midwinter famine.

The beaver we caught had to be treated respectfully. They fed us, giving us their sweet flesh. We would take their bones and tie them in pretty bags and carefully hang them in trees, so the beaver of the world would not feel the bite of dogs or foxes, and thus their spirits would not be offended.

The winter was the coldest in memory; even old Adaaweikwe could remember none worse. One day a stranger approached our camp. His name was Auginaush, and he came from the Wolverine River country to the north. Three people at their camp had died from hunger, and he said it was much worse at the other camps. One of those that perished at his camp was a girl I had met last summer. She was of my age, and I vowed to remember her when it was time for the spring memorial ceremony.

Later father said that the lack of meat during our hungry time may not have been all Gayaazhk's fault. The famine seemed to be widespread, and many others were having it worse than we did. He wondered what the people had done to so offend the spirits.

Birch saplings with beaver hides stretched on them to dry in the cold air.

The cold continued, day after day. We children grew tired and irritable. Our lodge seemed small and cramped, and we all did our work sullenly. The baby cried too much, my two new sisters bickered constantly, and I grew tired of them.

One evening, at a time when it seemed as if I could bear this place no longer, father called from outside, "Who is cold? Who is tired? For whom has the winter been too long?"

"We!" the children shrieked. "All of us," we cried, as mother and Adaaweikwe smiled. I smiled too, for I had forgotten what father always did this time of year. I followed the rest of the children outside, and father handed bows and arrows to each of us.

"Then shoot the Wintermaker!" he cried. "Shoot him from the sky."

Small arrows flew upward toward the great shimmering star figure, arms outstretched, sash shining brightly in the night sky. He, who had tested us so, was truly an awesome sight. We never saw where the arrows flew, but father assured us that our aim was true and that the Wintermaker was mortally wounded and would slowly fall out of the sky. The younger children laughed for the first time in days.

As usual, father was right. The next day the wind felt warmer, and in the evenings that followed we watched the Wintermaker tilt westward, toward the land of death. Slowly he fell, sliding forward, passing from the land. A great star figure defeated by us children.

As the warmth of spring came, so did the time to rendezvous with the fur traders. The furs we had collected all winter long were piled high on toboggans and would be traded for the items we needed and wanted. To make certain we were treated fairly, Adaaweikwe, who did not earn her name for nothing, went with father and uncle by dog team to the lake chosen for the rendezvous.

"Then shoot the Wintermaker!" he cried.

They returned two weeks later, just as we were packing up our camp, getting ready to head south to be reunited with our brothers and sisters at the sugarbush. Father said that this time the trader was new, a Métis, a person of scorched wood, and he had his wife with him, and she was related to mother's family. They both spoke perfect Ojibwe, so the trading was easy and quick.

Our beaver, muskrat, mink, fisher, and martin had obtained five new striped blankets of brilliant colors, the kind of cloth that keeps you warm even when it is wet. There were also large bolts of flowered patterned cloth, light enough for fine summer shirts; hanks of beads; two new muskets, with plenty of powder and bars of lead; and an assortment of other goods from files, knives, and axes to needles and thread. Finally there was a great iron kettle, which we would use at the sugarbush to make maple sugar. Father said the new trader would buy all the sugar we wanted to sell.

By the time we reached the sugarbush camp the sap had started to run, and the place was a whirlwind of activity. Great clouds of smoke and steam rose from the many fires, and a hundred people or more carried wood and sap buckets or tended the fires. The air yet had a chill to it, but the sun was warm, and we made much maple sugar in our new kettle.

It was here, during this time, that my first menses began, and my relatives built a small lodge for me. Mother and auntie and Adaaweikwe bathed me, rubbing me down with cedar, and combed my hair; they also brought me food each day. Once a strange man approached and thrust his hand into my lodge. There was a hideous sore on it, and I recoiled in shock. The hand remained, and finally I found the courage to touch it. The man then withdrew his hand without a word. Later he thanked me with a gift, showing me his hand, which had healed perfectly. It was then auntie thought I might have the gift of healing.

It was in the sugarbush camp that I again saw Gayaazhk, although he told me that was no longer his name. It was the first time I had seen him since he had left our camp to atone for his sin. That awful boy not only had survived the winter but was now tall and confident, and not one boastful word did he say. He gazed at me with respect and much admiration, and he began to bring my parents gifts of meat.

What a wonderful time that was. I knew I was living a blessed life, but not for much longer.

As I have already said, I try not to speak of the terrible times; it is enough to say that during the summer the spotted sickness came.

Auntie died first, then father. My father, good strong father, who strode through the snowy forest running down moose, and who shot straight and true, coughed his life away, his face hideously scarred. Adaaweikwe, my two cousins who spent the winter with me, and the one I was so much like, my mother—all gone. And so was Gayaazhk, boastful and teasing, then beautiful and strong, gone, the strength draining out of him like blood from a wounded moose. Even baby Anung, whom mother had called a little twinkling star, was now a star in the spirit world. How I hoped father or auntie was there when she passed over, so they could paddle her down the river of souls.

All the years gone by. So many dead, so many empty lodges. I never married. How could I, with no family in which a young man could reside, or which he could hunt for. Our ancestral trapping grounds are now hunted by someone else, and I live with relatives I barely knew I had.

Now an old woman, I stand here in the cold with the great Wintermaker blessing the land with snow and ice. The northern lights flicker, washing across the sky to the beat of an unheard drum. And my father, best in the dance, is dancing up there in his bright blanket coat, and everyone else—auntie, my two cousins, and my mother, in her black winter dress—is there with him. They all are wearing beads and colorful ribbons, and they bend and weave to the drum beat, clapping and chanting.

The river of souls, that sparkling, shimmering, promised river, will soon feel the stroke of my paddle, and then I will dance the next dance with father.

Earth Walks emerged from the darkness to stand beside her. Taking both her hands in his, he looked toward the audience. "My mother," he announced. "Moskobinedikwe is her name, a tribal elder and a wonderful storyteller. She puts me to shame!" And he gave her a long hug.

And my father, best in the dance, is dancing up there in his bright blanket coat, and everyone else is there with him.

To loud applause she bowed to the audience, then gracefully made her way off stage.

Watching her depart, Earth Walks said, "This was indeed a cold, harsh time in the planet's history, and winter was the coldest and harshest of the four seasons. For the Ojibwe though, winter was part of the ebb and flow of Mother Earth. In fact, even during the Little Ice Age many Ojibwe considered winter the best of the four seasons.

"There were many and varied reasons for this, but, given our time constraints, I believe the most important was the fact winter gave them what many of us complain we don't have nearly enough of: quality time with our immediate family, especially our children. Time to educate, train, tell stories, and just do things together. Time, through thick and thin, through normal winters and Mount Tambora winters, to create, maintain, and strengthen family ties and all the variables that go into them.

"Given our technology, I doubt many of us will ever have to face or live with the kind of conditions the Ojibwe did during the Little Ice Age. I also doubt many of us will ever experience their closeness of family, or the time they had to enjoy such closeness. Isn't the trade off between technology and life strange?

"I thank you for listening and watching, but I especially thank you for your patience. Remember, tomorrow bright and early, the buses leave from the rear parking lot at eight o'clock. Coffee, juice and doughnuts will be served on the bus."

PART THREE

THE FUR TRADER

The tour of the fur post was over, and we were all standing around in front of the main trading post, nosily chatting back and forth, waiting for the buses to return. Without the slightest warning the front door to the trading post crashed open, and out strode Earth Walks.

I had known he was going to do this, but I had no idea how he had planned to dress. So, like everyone else, I became quiet and stared at him in amazement, wondering where on earth he came up with this stuff.

Earth Walks wore a high-collared white shirt with lace cuffs and black buttons, a red vest complete with gold watch chain, and over these a black long-tailed coat right out of the early 1800s. His trousers were plain and black, he had dark, leather boots on his feet, and on his head sat a magnificent beaver hat.

DARBY (EARTH WALKS): My name is Simon Alexander Darby. I run this post, and I am also a voting partner in the North West Fur Company.

(Coming to the edge of the log porch Earth Walks sets his hat on a nearby chair and wipes his brow with a lace handkerchief that had been stuck in his coat pocket.)

When I came here to establish this post for the North West Fur Company I planned on staying two, maybe three years— suddenly I'm surprised to find that twenty have slipped by. If we exclude growing older and wiser, a lot changes in twenty years. For example, there's the way I dress.

The fine clothes you see me wearing today, I once wore every day. These are the clothes of a gentleman and meant for the streets of Montreal and London, or even for the Great Hall at Grand Portage. Now I wear these sorts of clothes only when receiving fine ladies and gentlemen like yourselves, or when I attend the annual partners' meeting at Grand Portage.

There is, however, one thing that has not changed too much over the years. If you have read any of the volumes of literature on the fur trade, or the voyageurs, or the opening up of this country, then you will have read how Native Americans became dependent on us traders for such basic necessities as clothing, cooking pots and utensils, and guns, and as some historians of your time have suggested, for food and shelter.

With your forbearance, and as I have a captive audience, I would like to tell you a little about myself, the construction and evolution of this post, how I conduct the fur trade, and what life is like here in the middle of the North American wilderness. As I tell you my story, you need to keep in mind the climatic conditions of this time, and the fact we traders and our voyageurs were once strangers in a strange and different land. When I have finished, you can decide for yourselves whether the concept of Native American dependence is correct, or whether your learned historians have once again grabbed hold of the wrong end of the stick!

To begin: I was born in Edinburgh, Scotland, and as a lad I loved the open hills and woodlands. I dreamed of having great adventures, traveling to new places, and, of course, making my fame and fortune. I soon realized the easiest and most prudent way for me to do this was to use my family connections in the fur trade business. I did just that and secured a position as a clerk's apprentice in Montreal. There I worked hard for three years, until I was eventually noticed and given a position of increased importance at the great fur center and rendezvous post in Sault Ste. Marie. Three more years of hard work, long days that turned into even longer nights, and then along came the opportunity I've been waiting for. I was put in charge of a new fur post, the very one you are now visiting. It was my job to build it from the ground up, and if successful, I knew I would pretty much have it made.

So late that summer, loaded with trade goods and accompanied by a dozen voyageurs and a few laborers, I set off along Lake Superior to Grand Portage, then south to where we stand today. Here I was to winter over, and in the spring conduct the business of the North West Fur Company.

The journey from Sault Ste. Marie took several weeks. I traveled by birchbark canoe, which was built by, and purchased from, the Indians. In fact, if it wasn't for Native American birchbark canoes, there would not have been a fur trade!

The journey was one of long days and short nights, and the voyageurs carried the trade goods over the many portages by using "tumplines," a device invented by the Indians.

When I started out I had planned to wear a pair of English-made boots, sturdy and dependable. However, on the advice of one of my voyageurs, who told me the boots would last about three weeks, I obtained a pair of Indian moccasins—much more comfortable and a lot more practical.

The air was turning crisp, the nights colder, and long lines of Canadian geese moved south when we arrived at our destination. The laborers immediately set to work building our quarters. I tried to supervise them as best I could, but I also had many other things to occupy my mind and my time.

The first thing I quickly found was that we were dependent on the local Indians for all of our food. So, as I was leader of the establishment, it was up to me to swap trade goods for a stock of maple sugar, dried and smoked whitefish, pemmican—which, thank goodness, was replaced by frozen moose, deer, and bear meat during the winter months—and wild rice. Garden produce like corn, beans, and squash also came from Indian women, and these completed our winter larder. I did bring with me seeds for potatoes, peas, and parsnips. These I gave to the Indians. They agreed to plant them in their gardens the next spring and share the harvest with me.

Late that summer, loaded with trade goods . . . I set off along Lake Superior.

The second thing I realized was that I would need tools for surviving the winter. I had to secure dog teams, dog harnesses, and toboggans from the Indians before the snow came. As one of my contemporaries at a post on Rainy Lake wrote, "I have no one here to make raquettes"—meaning snowshoes, because the English thought they looked like tennis rackets. "See what it is like to have no women nearby, try and get racquettes, as there is no moving without them." Without racquettes and the women to make them, he was out of business.

I also needed a warm winter capote, or coat, and this I had to order from an Indian seamstress who lived nearby. She would make the coat out of either a moose hide, or a fine English blanket.

Finally, I found I needed the Indians' geographic and geological knowledge to point out the locations of springs, local pathways, and the best canoe routes. They also told me the boundaries to their lands, where other bands lived, whether or not these other bands were amenable to trade, and what I must do to win them over. Any map-making or exploration would be done with their guidance and advice.

That first year it turned out we arrived too late to finish all of the necessary buildings. Though my crew of laborers worked hard, there was just too much to do. So I had to employ Indians to help finish the job. Indian women gathered bark for the roofs, collected the rocks for the chimneys, and showed me where the clay deposits were located so I could chink my logs.

Even as winter arrived I found myself thinking about spring. In this regard I had to secure a supplier of additional canoes. For this I went and spoke with one of the important elders in the band. This person gave me the name of the best canoe-making family. I then had to arrange for this family to make our canoes, agreeing to pay them in trade goods.

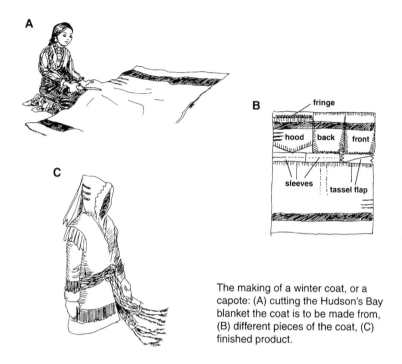

The making of a winter coat, or a capote: (A) cutting the Hudson's Bay blanket the coat is to be made from, (B) different pieces of the coat, (C) finished product.

To be a fur trader, I of course needed furs, and lots of them—for my supervisors, who seemed to care little for my living conditions, had quotas I must fill.

This meant I had to convince the local Indians to bring their furs to my post, and not take their business elsewhere. It turned out this was not as easy as I originally thought, for I found the Indians to be far more mobile than I, and in a position to go where they wanted, or where they could get the best deal. According to custom, I distributed what I considered to be fine gifts to the appropriate elders, and hoped they were impressed.

As the winter dragged on, I realized I was almost totally dependent on the Indians. They were the ones who trapped the animals without damaging the fur, they processed the skins, which had to be fleshed and dried, and, finally, they brought the hides to my post.

At last spring arrived, but for me it was not a time of celebration and joy—not yet. Instead, it was a time of worry and frustration. Would I get my quota of furs? Would I be in business next year? Would I still be a rising star in the North West Company?

It turned out my gifts and trade items were considered good, I happened to offer generous trade terms, and I managed to offend no one, and I respected the Indians' ways of doing things, so the furs poured in. My first year was a success.

With the coming of spring the winter's take had to be packed into ninety-pound bales. The voyageurs, with a few hired Indians amongst them, transported the furs back to the rendezvous post at LaPointe.

With the safe delivery of the furs, my first season came to an end, the post was firmly established, and, as it turned out, so was I. In this manner I began my second year.

However, this next year was much different. Not only had I learned a lot about the land, the Indians, and the fur trade, but I had done so well my employers wanted me to remain here, and offered me a 'winterer's' partnership in the company to do so. When I accepted, my new partners advised me to find a wife. In my own mind this was not such a bad idea, for not only would this cement relations with the Indians, but it would also alleviate the terrible loneliness I felt during the long winter past.

My chosen bride was the daughter of one of the leading families. The marriage was arranged, as is the custom in my culture as well as hers. My wife would teach me the local language, and her brothers, father, uncles, sisters, and aunts would be loyal to me as customers and providers.

I put together a great store of trade goods: calico cloth, ribbons, kettles, metal goods of all kinds, rich blankets of the brightest colors—all of this for a bride price. This is just the opposite of the customs in my country, where the bride's family pays the groom. The bride price I paid had to be commensurate with the skills of the bride, and to balance the loss of a talented daughter to her family. My wife is an excellent seamstress; she knows all the arts and skills required to live in this part of the world. She knows several languages, over two hundred different types of plants and how to use them, the varying quality of fur pelts, and the customs of bands and tribes in the surrounding region, and she handles a canoe as well as one of my voyageurs. She also brings her people and mine together for many different social events. On my part I am required to treat her well, earn her respect, and form a partnership with her, for she retains the right to divorce me, and that would ruin me and this fur post.

In my culture a marriage ceremony consists of an exchange of vows before witnesses; no official of church or state is required. So it is with the family of my bride. There was a brief ceremony, with her family in attendance. It was brief because, as I now know, funerals are greater events in the lives of these people than marriages.

As it turned out, I was not the only one being married. The voyageurs and laborers also took brides of their own, and, as well, some of them took in their brides' aging parents. As you have seen, the voyageurs and laborers have their own cabins, all in a row along the lakeshore. These were built in the French manner of land division, each having access to water and the forest behind.

Almost overnight, or so it seems to me, I had children of my own, and all too soon they were old enough to join the others at their different games. Watching them play makes me realize that not only have the years flown by, but somewhere in that time this post has turned into a community.

And I find I am also much changed. I am content, fairly well off, as a wintering partner should be, and I find my ambition to move onward and upward has been lost somewhere in the heart of this great land. My family grows, and soon my four children will take over the operation from me, becoming guides, traders, and interpreters. My Indian bride and I have grown older, and I now find I spend more time with her, especially in the evening when I go over the accounts, or read, and she does her beadwork, sewing, or mixing of herbs. And to my everlasting amazement, I realize I now converse in her language far more than my own.

If I had to answer the question I posed at the start of this, I would tell your historians that dependence was certainly a two-way portage. However, it seems to me that the essential traffic, the knowledge and basic necessities for surviving on this land, flowed mostly from the Indian to the European and American. And that's the way it was from 1570 to about 1820 in fur country.

Now that my story has arrived at the end of its journey, I can tell you the buses are waiting in the parking lot. I thank you for visiting my fur post and hope you enjoyed your time here. I bid you good voyage, and pray you have fine times on your journey home.

In the bus, on the way back to the auditorium for the final wrap up, I told Earth Walks he was a natural playing the part of a North West Company fur trader.

Smiling, he tipped an imaginary beaver hat in my direction. "I should be," he said. "I spent many afternoons listening to my grandfather do much the same thing. He was an interpreter at a restored fur post."

"And his name was?

"Darby, George Darby."

"So you really are a Darby!" I exclaimed.

"Absolutely. My given name is John Darby."

"And Earth Walks?" I asked.

"You could say it's my professional name. I learned many years ago that people are much more interested in hearing stories about Native Americans from someone named Earth Walks than they are from John Darby."

"If someone wanted to get in touch with you, whom would they call?" I asked. "John Darby or Earth Walks?"

"Earth Walks," he quickly replied. "I like the name and I have come to believe it suits me."

And that was something I couldn't argue with.

CHAPTER SEVEN

MAKERS OF THE MAGIC SMOKE

PART ONE

THE RED STONE

I hate early-morning phone calls. Waking in the weak light, fear causing a minor earthquake inside the stomach, and the necessity of lifting the receiver and having to be civil to hear the dreadful news.

"Hello," half mumbled, half screamed.

"You're home," a cheerful voice. "Back from exploiting Mother Earth, I presume."

"Earth Walks," I groaned, relief making me feel ten pounds lighter.

"Good to hear your voice too."

"It's four in the morning! Do you know that?" I yelled.

"When the spirit calls, follow her you must," he answered. "Listen. I apologize for the hour but I have been working on this

Pipestone story for what seems forever. I just about have it, but I need to go down to the quarries to check on some descriptions and take pictures. I also need geological advice.

"I was packing my stuff when a bolt of lightning struck me. I remembered your interest in the quarries and your moaning about never having had the chance to poke around in them. So I thought hey, why not fulfill your wish and at the same time get the help I need. Besides, I figured it would give us an opportunity to discuss the upcoming lecture program."

"It's four in the morning," I yelled again, my mind obviously in neutral.

"You keep saying that," he replied. "Sure I'm not talking to some sort of answering machine?"

"Man—I don't believe this!" I exclaimed. "Pipestone? Now? Are you crazy? What lecture program? What are you talking about?"

"Aaah," he softly purred. "No one called you?"

"I just got back in town. I haven't even been to my office."

"Ummm," he said, as if that explained everything. "This will be news. We have been asked to repeat our program. But not here. The muse calls us south to the big city. To educate and entertain larger audiences and, of course, receive bigger rewards.

"Listen," he quickly continued. "Think about it as you pack. Overnight will do, and bring a sleeping bag. I'll pick up coffee and doughnuts and be there in about ninety minutes."

Before I could voice the appropriate four-letter response, the line went dead.

"Why me!" I shouted at the ceiling.

The harsh sound of the dial tone was the only answer.

Hanging the phone up, I shook my head in resignation and went down to make coffee.

I was halfway through the first cup when I made up my mind. The early hour notwithstanding, I did want to see Pipestone, a place Native Americans consider sacred ground. It was also a place where prehistoric people, for thousands of years, had used and practiced geology. I also knew if I went with Earth Walks I would get a much different perspective on the quarries than the usual tourist. Finally, I had to admit to myself, it would be good, as well as challenging, to see Earth Walks again. My decision was made. I would go to Pipestone before the sun came up.

Caffeine flowing through my veins, shaved, and packed and waiting for Earth Walks, I pulled out *Minnesota's Geology* by Ojakangas and Matsch. Sitting at the kitchen table, so I could see the driveway, I reread the sections on pipestone—quarry and rock—and the Sioux Quartzite, the metamorphic rock that underlies most of the Pipestone area.

I had just closed the book and started on my third cup of coffee when Earth Walks pulled into the driveway. It was a quarter to six.

He greeted me with an enthusiastic bear hug. Throwing one arm around my shoulder, he swept up my backpack and ushered me out to his Subaru Outback. "Time heads west with the coming sun," he said, throwing my stuff in the back and settling into his seat. "Let's go catch some. Coffee's in the bag, doughnuts in the box on the backseat. I'm starved."

On our way to southwestern Minnesota and Pipestone, Earth Walks and I caught up on the two months since the lecture series had ended. As we talked we ate half-a-dozen doughnuts and drank too much coffee. I think we visited every rest stop between my house and Pipestone. We also discussed the possibility of collaborating on another lecture program.

It was during this discussion that Earth Walks told me, "The story I am working on concerns prehistoric Native Americans who quarried at Pipestone, and how the Pipestone area came to be considered common ground, a place where all Native Americans could meet, live, and work in peace.

"If you could provide the geological background for the quarries and my story, I think we would have a solid addition to our program."

"It's possible," I told him. "It all depends on what kind of cultural and historical material you plan on talking about, and whether or not the geology will blend with it."

"Ah," he said, "You want to know what I've come up with. For that I have to organize," and he fell quiet, seeming to concentrate solely on his driving.

I settled back in my seat and watched the farm fields of central Minnesota roll by as I tried to recall what I knew about the geology of the Pipestone area.

"Do you know the poem 'Song Of Hiawatha'?" Earth Walks suddenly asked.

"Not really," I replied, snapping out of my geological reverie. "Hiawatha has something to do with Pipestone?"

"Scientists," he grumbled, and sadly shook his head. Gripping the steering wheel with both hands, he threw his head back and thundered,

> From the red stone of the quarry,
> With his hand he broke a fragment,
> Molded it into a pipe-head,
> Shaped and fashioned it with figures;
> From the margin of the river
> Took a long reed for a pipe-stem;
> Filling the pipe bowl with tobacco,
> Gitche Manito, the mighty,
> Smoked the calumet, the peace-pipe,
> As a signal to the nations.

"Or something like that. Anyway, that will be my introduction. After which I plan to use the line about 'a signal to the nations' to point out that Native Americans did indeed come to Pipestone. They came by boat up the Mississippi and then along the sluggish Minnesota, or by way of the rolling Missouri and a tributary called the Big Sioux. They also came by foot from the plains and the woodlands; there were buffalo hunters, elk and deer slayers, maize and bean planters. The people of the Indian nations came to southwestern Minnesota to quarry and take home a soft, red rock called pipestone. And they kept coming for more than 3,000 years."

"The place these people were paddling or hitchhiking to is called the Coteau des Prairies," I said, as easily and quickly as if we were back on the lecture stage.

"Highland of the prairie?" he asked.

"You got it," I answered. "A flatiron-shaped plateau that descends, in a series of wide steps, to the Missouri River on the west and the Minnesota River on the east. The plateau is some five hundred to eight hundred feet higher than the glacial deposits of the surrounding countryside. It stands tall and proud because it is made out of a relatively erosion-proof rock called the Sioux Quartzite—a metamorphosed sedimentary rock, originally sandstone."

"See!" he exclaimed. "That's exactly why I need you. That's perfect.

"Did you know prehistoric Native Americans traded pipestone across a good part of this continent? The Anasazi people, the mound-building Hopewells in Illinois and Ohio, and the builders of Cahokia traded for and used the red stone. These people, and most other Native Americans, used it primarily for the making of pipes, ceremonial pipes or calumets, among which the most important was the pipe of peace."

"And all this red stone came from the Pipestone quarries?" I questioned.

"Not all, but most," he answered. "And the reasons were both practical and spiritual. On the practical side, the red stone was said to be more easily carved and made a better pipe than any other stone. It was so soft it could easily be carved with stone and copper tools, and the pipe bowls and shanks could be drilled out with flint, quartzite, or copper points. These points were attached to a hollow reed that was rotated between the palms of the hands.

The red stone was used primarily for the making of ceremonial pipes or calumets.

The finished pipe is a work of art, with incredible beauty and wonderful craftsmanship.

"The stone they carved, and still carve, is beautiful. The colors range from scarlet and vermilion to blood red, and more rarely spotted stones, red with cream and pinkish spots. When the finished pipe is polished by rubbing it with buffalo fat or a similar substance, it is a work of art, with incredible beauty and wonderful craftsmanship. Wait until you see some of the finished pipes. They will take your breath away.

"On the spiritual side," he continued, as he stuck on his sunglasses against the glare of the rising sun, "the quarries where pipestone was and is mined have tremendous religious and symbolic importance for Native Americans. The pipes carved from this stone are integral parts of ceremonies and customs, for these are sacred pipes, representing a means of sending prayers and thoughts to the creator above, to Mother Earth below, and of honoring the four sacred directions. If a prayer or thought is uttered as smoke from the pipe is blown into the air, then this prayer or thought will be carried by the rising smoke to the creator, or, if it is blown downwards, directly to Mother Earth."

"Hence the story of White Buffalo Woman?"

"Exactly," he answered. "There are many stories and myths associated with the quarries and with the sacred red stone."

"That I can imagine," I responded. "But you know, geologically speaking, the red stone is nothing more than hardened clay. In fact it is made up of the common elements aluminum and silica, and the red color is due to the mineral hematite, or red ocher, an iron oxide. This iron-rich mineral gives the rock its red color because it has been thoroughly mixed with the clay, like strawberries with ice cream to make a milkshake. The red stone occurs as thin layers, only sixteen to twenty inches thick, between thicker slabs of the much harder Sioux Quartzite. This arrangement is much like cheese slices placed between thick chunks of French bread."

"And how do geologists figure the pipestone formed?" Earth Walks asked.

"Both of these rocks," I answered, "the thick and the thin, were once sedimentary rocks that formed on the bottom of a shallow sea some 1.4 to 1.6 billion years ago. In a sense both are fossils, for they represent the remains of a range of lofty mountains that covered the Lake Superior region some 1.6 to 1.9 billion years ago. The erosion of these mountains, over hundreds of millions of years, produced abundant sand-size grains of quartz that were carried by streams and rivers to the ancient sea.

"It was during dry times, when water levels were low and stream velocities too slow to carry or transport the sand grains, that the clay-size material got its chance. Being lighter and finer than the sand, the clay could float, like a toy ship, down the slow-moving current all the way to the sea. Once there, it would slowly settle to the sea floor, like snow flakes falling to the bottom of a snow globe, to form the layers that would one day become your sacred red stone. Slow times in these rivers were few and far between, which is why the clay layers are thinner and rarer than those of the quartz-rich sands. Over time the millions upon millions of sand grains were deposited and cemented together, to form a rock called sandstone.

"With a rather large dose of geologic time," I continued, now definitely on a geological roll, "the sand and clay deposits were buried deep in the earth. This burial caused a metamorphosis to take place; like a caterpillar turned into a beautiful butterfly, planet earth changed the weak, friable sandstone into a hard, tough metamorphic rock called quartzite, and the sticky red clay was hardened into carvable pipestone, which geologists call mudstone, or argillite."

"Excellent!" cried Earth Walks, almost driving off the side of the road. "Simply wonderful," he bubbled. "You realize this gives us a marvelous opportunity."

"For what?"

"In the past we have taken Native American stories, oral history, and tried to place them in the context and confines of geological events and processes. Here we have the opportunity of taking your geological story and showing how Native Americans might interpret it."

"Interpret it?" I questioned. "What do you mean by that?"

"Let me show you," he said. "It could be said that the parents of the red stone—your aluminum, silica, and iron—once sat high on lofty mountains with Gichi Manidoo. Then with the great flood, when the whole earth was covered with water, they were washed to the bosom of Mother Earth. There they were reborn as the sacred red rock of the pipestone quarries. It is from this red rock that the creator then made the ancestors of all the Indian people."

"Not too shabby," I said. "You've obviously been giving this a lot of time and thought."

"That I have," he agreed. "And, to continue, since many Native Americans came to believe the first Indians were created from the red rock, pipestone came to be considered part of their flesh, or to represent the flesh and blood of their ancestors.

"Out of this belief grew many legends. One of the oldest and best known is the one made famous by Longfellow's poem. This particular legend claims that after the American Indians were created and had spread as different bands across the continent, the Great Spirit called them back to the pipestone quarries. He did this by breaking off a piece of the red stone and molding it into a large pipe, which he filled with tobacco and smoked. As he smoked, he pointed the pipe in succession to the four sacred directions, north, east, south, and west, and on his final puff, he honored Mother Earth. When the people had arrived, a representative from each tribe, he told them that they had been molded from the red stone, and that they should take it, make pipes out of it, and use these pipes as a means for finding peace amongst themselves.

"And so the pipestone quarries became sacred and neutral ground, held and owned in common by all native people. A place where all Native Americans could come and quarry the stone and so renew their sacred calumets. When the Native Americans were much more abundant than they are today, the quarries may have been a noisy, busy place."

As I listened I watched the Coteau des Prairies rise up out of the glacial moraine we were driving over and, in a series of gentle swells, begin its slow climb toward the sky.

When Earth Walks finished speaking, I said, "Tell me, your story about the quarries, when does it take place?"

"Summer. About one thousand years ago," he replied. "Why?'

"Your ancient quarriers wouldn't recognize this place," I said. "It's nothing but paved roads, ribbons of blacktop, that cut across one cultivated field after another, all the way to the horizon. Prehistoric Native Americans would think they had been transported to another world."

"And they would have been. They would be strangers in a very strange and different land. Did you ever hear of a man named George Catlin?"

"As in catlinite?" I asked.

"The very one. It was Europeans who named the red pipestone catlinite in his honor."

"And honor is all that it is. The word has absolutely no geological meaning or significance."

"Fame is so fleeting," Earth Walks commented. "Anyway, Catlin crossed the highlands in the 1830s and wrote that the area was empty of anything that grew except the grass and the animals that walked upon it. Later, around 1860, an explorer by the name of Fred Ceavenworth described the highlands as swell after swell without trees, bushes, or rocks, and he said it was everywhere covered with green grasses, giving the traveler, from its highest point, the most open view of nothing save the boundless ocean of prairie that lay beneath and all around.

"How do you remember this kind of stuff?" I asked, being just a bit jealous. I couldn't remember passages or phrases like that for more than a couple of hours after I read them.

"Just read it," he replied, shrugging his shoulders. "If it sounds interesting I tuck it away in the old filing cabinet," and he tapped his head. "This gift of being able to recall what I read certainly helped me get my anthropology degrees."

"You never really used them, did you?"

"I use them everyday," he sharply responded. "Though it's probably not in the sense I think you mean. Even though I've been a tenured university professor, those degrees, for me, provided the necessary training and background that have allowed me to battle for my people's history, my history, and do so on a level field of credentials and pedigrees. And I can tell you it's been a tough war. It was this constant fight for recognition of Native American's holistic worldview of planet earth that has led me to do what I now do."

"Must be your true calling," I told him. "You are really good at it."

"Thank you," he said with an agreeable nod. "Sometimes I actually believe I'm blessed. But those moments are fleeting. But back to Catlin and Ceavenworth and a time the prairie grasses were much different than they are today. Many of them would have stood higher than this car."

"And it's all gone," I said. "Cut down, burned out, plowed under to make way for corn and wheat and who knows what else."

"Not quite all. Don't forget Pipestone. It is managed by the Feds to recreate as much of the historic tall grass prairie as possible."

After that we fell quiet and watched the green and yellow fields that now made up the Coteau des Prairies unfold before us. A half hour later we arrived at the Pipestone National Monument.

The tour of the monument was as good as I expected, and the quarries were fascinating. But the best part for me was being able to spend a few hours at an active quarry with a friend of Earth Walks.

Earth Walks, telling me this would help make up for the liberal arts education I had missed out on, went off to check the accuracy of some of his descriptive passages and to take some detailed photographs.

I watched with admiration the well-muscled Native American work at breaking his way down through the Sioux Quartzite, inch by backbreaking inch, to reach the sacred red rock. Later I helped carry broken pieces of the quartzite out of his pit. As he worked away, switching from sledgehammer to crowbar, then hammer and chisel, he told me the legislation that established Pipestone as a national monument also stated that the quarrying of the red pipestone was reserved to Indians of all tribes. Today, Native Americans from many different tribes still come to the monument to quarry the sacred stone.

I also found out today's quarriers are modern in name only. National Park Service regulations state that the stone may be quarried only with the use of historic kinds of tools and methods. The term "historic," as used by the National Park Service, does not mean the hammer-stones, flint, and copper that the prehistoric quarriers used, but tools from the mid to late 1800s: iron and steel tools such as crowbars, sledges, shovels, pickaxes, and wedges.

Even with these longer-lasting, sharper tools, the work of removing the Sioux Quartzite to reach the pipestone remains difficult and hard. Today's workers, however, are able to quarry down to depths of ten feet and to advance their pits twenty feet or more in the direction the pipestone dips or slants. These are depths and distances the prehistoric quarriers would marvel at.

Today's quarriers also strive to keep alive the traditions that surround the red stone and make it so special. Many of today's workers, like the man I was with, conduct special ceremonies before they begin to quarry. These range from offerings left at the sacred rocks (the three maidens) to purification ceremonies in sweat lodges and other traditional ceremonies honoring Mother Earth and the Creator. They also carve pipes for ceremonial use, ceremonies that are intended to preserve and continue the heritage and culture of Native Americans, regardless of tribe. And like the early quarriers, these traditionalists believe a pipe carved from the stone is not sacred until combined with a proper stem and smoked in a traditional and sacred ceremony. Otherwise the red stone is but a beautiful rock, something that, with the blessings of the spirits, can be turned into a fine work of art, a work that may express Indian traditions and signify a way of living that has passed from the land—a way that may be seen, shared, and learned through the art of the pipemaker.

PART TWO

COYOTE AND THE GIFT

A week after my visit to Pipestone, a large box arrived courtesy of UPS. It was from Earth Walks, and inside were two separate packages. One was a manila envelope with "Pipestone" written across the outside, while the other was a box about the size of a shoebox. On it was written "For your early-hour sacrifice and the pleasure of your company at Pipestone."

Opening the envelope, I found inside a thin manuscript and a letter from Earth Walks. The letter read:

"Enclosed is the Pipestone story. Read it at your leisure (or better ASAP) and let me know what you think, especially the idea of adding this to the program.

"As you read the story remember its context and the background info the audience will already have. Any geological mistakes or misrepresentations please feel free to correct— Mother Earth and I won't mind one bit!"

Setting the letter aside, I opened the second package and found myself looking at a seventeen-year-old bottle of Ardbeg Scotch. For someone who knew nothing about single malt Scotch, Earth Walks had good taste!

That evening, glass of Scotch in hand, background notes beside me, I settled down in my favorite chair and read Earth Walks' new work. He called his story "Coyote and the Gift." What follows is Earth Walks story, just as I received it.

Scene I: *The set consists of three large panels. Panel one is a picture of a tall grass prairie; panel two is the same prairie but dotted with rock outcrops, and the end of a rocky ledge of Sioux Quartzite is seen in the distance; panel three is a view of the rocky ledge, waterfall, pool, and creek as they occur at Pipestone. A Native American camp, with two very different lodges, is set up in front of the first panel; three grey boulders, representing the three maidens, sit in front of panel two; and an active quarry is set up in front of panel three. As the scenes change from one setting to the next, the active area will be lit by spotlights.*

NARRATOR: There is a tradition spoken of in the legends of the pipestone quarries. In some tribes there were said to be men who were experts at quarrying the red stone, men who were also master craftsmen, carving and shaping the stone into beautiful works of art. These men were believed to have been touched by the Great Spirit, or blessed by Mother Earth; these were special men, much honored and respected. Among them was a tradition of carving their best and most beautiful pipes as gifts to their sons when they went through the tribal rituals of manhood.

Sit back now and try to imagine what it might have been like to be such a man; a man who quarried with flint, chert, and copper tools, who traveled tens to hundreds of miles to claim a special piece of rock from Mother Earth, a rock from which a very special pipe could be carved.

The following dramatization is called "Coyote and the Gift" and takes place during summer time at the Pipestone quarries some nine hundred years before the present.

(Earth Walks, as an older Wakinyantawa, sits on one side of the stage with several children around him. He is an old man telling the children the story of one special summer at the pipestone quarries. Actors actively portray the events Earth Walk's narrates.)

A spotlight illuminates Earth Walks and the children. Earth Walks pulls a buffalo robe tighter around his bent shoulders.)

WAKINYANTAWA (EARTH WALKS): I tell you children this story so you may know something of the place our sacred pipes come from, and something of holy ground and the country of spirits. I also tell this story so you know more about the men who travel far from home to quarry the sacred rock. Finally I tell you this so you remember to be alert and never take anything for granted.

Looking down what we call the valley of time, it seems to me just a summer or two ago I last went to the sacred quarries, not the thirty-two that have passed me by. Strange, out of all those summers, how I remember one more clearly than all the others. Not like you children, who remember only the one you now race through!

In the summer I tell about, I arrived in the sacred valley during the quiet time that comes with the rising of the sun. The perfect time for the spirits to hear the prayers and thoughts of Wakinyantawa. The perfect time for making my offering to the guardians of the quarries. *(A young man, portraying a younger Wakinyantawa, enters and walks over to stand before the three stones that represent the three glacial erratics called the three maidens.)*

I clearly remember standing in front of the stone guardians and carefully unfolding a deerskin pouch. Out of this I took a polished piece of the red stone. The piece was about the size of my hand, and I had spent several cold winter days working it into the shape of a swimming turtle. Where the shell of the turtle should have been, I had carved a shallow bowl, and this I carefully filled with my finest tobacco.

Lifting the turtle above my head, so it shone brightly in the first rays of the coming sun, I asked the guardians to grant me good quarrying, to give permission to take the red stone away so it could be shaped into sacred pipes, and, last, to permit me to quarry a most beautiful piece of red rock. Out of this I would fashion my finest pipe, a ceremonial pipe to be presented to my son on the day he became a man.

Placing the red turtle at the base of the largest of the three rocks, I knelt and uttered my final prayers.

I firmly fixed my tool sack on my back, picked up my walking stick, and strode down the small slope into the valley of the Great Spirit.

The small valley is the home of the quarries. The valley begins at the edge of the three sacred stones and ends some 1,200 paces away at the foot of a solid rock wall. Here, I will draw you a picture so you can see it more clearly.

The wall, which is six or seven times my height, is made out of an evil pink rock that is harder than pemmican. On each end it slopes gently, like this, into the flatness of the grassland.

Out of the shadows of the wall you can hear the busy sound of falling water. There are many small springs behind and along the top of the wall, and these feed a small stream, which slowly gathers itself into a steady flow before rushing over the steep edge. The water plunges in long, silvery sheets into a green pool that has, over the ages, been worn into the hard, pink rock. As you children can imagine, this pool was a welcome place at the end of a long, hot day.

It has been said the Creator himself stood upon that very wall, at the exact place where the stream rushes over the edge. From this spot Creator reached into the valley and scooped up a handful of the red rock. Saying the rock was made of the flesh of all the Indian people, he fashioned a great pipe, smoked it in the four sacred directions, and then blessed Mother Earth with it. In doing so he made the ground sacred, a place where all who were of the red rock could come in peace, quarry the red stone, and take it home to be carved into sacred pipes, pipes for prayer, friendship, and peace.

The Creator, when he filled his hand with the red rock, left big chunks and round knobs sticking out of the ground over the entire length of the valley. These bits and pieces are tilted at an angle to the ground, so they slant down, beneath the valley in the direction of the rock wall.

I wished Creator had taken more of the red rock, and in doing so left many more chunks sticking out of the ground. This would have made my life a whole lot easier. As it was, all of the good stone had been taken from the exposed rocks long ago. This meant I and all the other quarriers had to open new pits, and to do this we were forced to move closer to the wall. And closer to the wall meant we had to break our way down, through layer upon layer of the hard pink stone, in order to reach what we came for. And it was long, backbreaking work, day after day, laboring from sunup until sundown in the hopes of having the ancient saying proved true;

> The harder the labor,
> The thicker the blood,
> The richer the stone.

Two seasons before the one I tell about, I made my furthest advance, moving to a place halfway between the red rock knobs and the waterfall. I selected the spot because it was here, on a gray, misty morning, I had seen a bushy-tailed fox take down not one, but two fat hares. I was sure this was a sign from the spirits that the red rock lay thick and full beneath that very spot.

I spent a whole day digging through the yellow soil that is beneath the tall prairie grass. Then more than two complete circles of the moon smashing, wedging, prying, and cutting my way down through the stubborn pink rock. And stubborn it was, for I went through enough hammer-stones to fill a beach—chipped, splintered, and broken, all to make cracks wide enough for a stone wedge or a wooden pry bar. Once I pried and wedged a piece loose, I had to break it into smaller pieces so I could carry it out of the deepening pit.

But by the Great Panther, in the end it was worth it. When the red rock was finally exposed I fell to my knees and thanked Creator and the three guardians. Then I wept like a baby. For the sacred stone was not only full and thick, as the signs had foretold, but its color was like none I had ever seen. It was blood red, with flame-shaped streaks of cream, pink, and pale purple shot through it, and in places it was covered by small white and red spots.

See, all of you, look, here is a piece I have kept, raw and unpolished, a magnificent stone, easy to carve and shape; and the colors of the finished pipes give them a beauty greater than a rainbow on a wet, summer morning. Look for yourselves. And here, look at this, a pipe I carved from what came to be called the rainbow rock, a rock carvers clear to the shining mountains would trade their teeth for.

The next season was one of my best. I collected enough red stone for all the sacred pipes I would need to make, and many fine pieces for trade, and all in less than two moons. This left me plenty of time to begin the new advance of the pit, outward and downward, to expose a new layer of the rainbow rock. In the time before the Wintermaker, I managed to quarry halfway to my goal.

Here is a piece I have kept, raw and unpolished.

The next summer, the summer I now tell about, I wasn't the first to arrive at the quarries. As I made my way across the sacred valley, I could hear a steady hammering coming from somewhere close to the rock wall. It was tradition among the stoneworkers for a new arrival to go and visit all the other working quarries. The purpose was to announce your presence and together smoke the pipe of peace. This being the custom and the working way, I placed my tool sack in my quarry and prepared myself to do this.

Scene II: *A working quarry with broken rock piled around one side. In the quarry, hard at work, are two Native Americans. Three new panels (made by reversing the first set). Panel one—a picture of a tall grass prairie; panel two—the prairie, a few rock outcrops, and several quarries; panel three—the rock ledge, waterfall, and pool, only now the viewer is much closer than in the first scene.*

WAKINYANTAWA (EARTH WALKS): I followed the sounds of the hammering. Soon I came to the only active quarry in the valley. This was a recent quarry. It had been started two years ago and abandoned soon after. The pit was located closer to the rock wall than my own, and once the original workers found out how far down they had to go to reach the red stone, they gave up and left. The current occupants had greatly enlarged and deepened the quarry, meaning they must have arrived soon after Old Man melted away.

Standing on a large pile of rock, which had recently been carried from the pit, I watched one worker pound away at the hard rock, while his companion used a flint wedge and stone hammer to break apart a large, loose piece.

The two quarrymen were of medium height, thin and well-muscled like myself, but there the similarities ended. These two had short hair, and it was tied in a round knot at the back of the head by a piece of brightly colored material. Their skin was paler than mine, and their leggings and shirts were made out of a beautiful green and white substance that seemed to mold itself to their bodies. The way they worked showed they had quarried stone before.

Watching them, I opened my deerskin pipe bag and withdrew my quarry pipe. This pipe I used only in the sacred valley. On one side of the pipe bowl I had carved a human figure holding a hammer-stone, on the other side one of the sacred rocks, and a thunderbird, its wings folded behind it, sitting at the rock's top. Attached to the stem was an eagle feather, a piece of turtle shell, some of the fur of the nosy fox, and a small piece of the sacred rock, all of which gave this pipe much power when used within the sacred valley.

Filling the bowl with tobacco, I held the pipe toward the two workers, and loudly called;

YOUNG WAKINYANTAWA: Brothers, I am Wakinyantawa from the country of the mighty buffalo, the land of tall, rolling grasses. I come to renew the sacred pipes of my people, and I come in peace and friendship. I wish you good quarrying, and pray the spirits smile on us all.

WAKINYANTAWA (EARTH WALKS): It was only later I realized the sound of my voice, loud in a place where they thought they were alone, had scared them as much as the wail of a spirit walker. The one using the hammer-stone dropped it like it was on fire. Throwing his head back, he stared at me as if I were an eagle and he a trapped gopher. The other was frozen in time, just as if he had been bitten by a medicine snake.

Seeing the outstretched pipe, they both relaxed and started to smile. The one holding the stone wedge took two steps toward me, raised his arms in the air, and began to speak.

As far as I was concerned, he could have been speaking to the rocks, the clouds, or the sky, for I understood not a single word. Shrugging my shoulders, I replied that I did not understand and, pointing to my pipe, I made a circle with my hand.

Nodding and smiling, they drew circles in the air with their fingers. Slowly they climbed out of the quarry and joined me at the top. When they were sitting, one on either side of me, I lit the pipe and took a long puff. Blowing smoke toward Mother Earth, I passed the pipe to the one on the left, who took a deep puff and blew smoke high in the air, toward Creator. The pipe was then handed to his companion, who drew deeply and blew a cloud of smoke down, toward Mother Earth. The sacred pipe was then passed back to me, and I took one last puff, blowing the smoke toward the Creator, and completing the circle.

When this was done the one who had spoken to me pointed to himself and said, "Kalsatake," turning, he pointed to his companion and said, "Situakee."

"Wakinyantawa," I replied, pointing to myself.

"Wakinyantawa," they repeated, and then, with more signs, made it clear I was to camp with them, which was the custom, and, when the sun sat on the edge of the world, share the evening meal.

Signing I understood, I pointed in the direction of my quarry and made the motions of breaking rock with a hammer-stone.

Nodding they understood, they gave me a final smile, then scrambled down into their pit. Without another look in my direction, they took up where they had left off.

So began the days of hard labor, days of sweat, backaches, broken tools, and stiff muscles. What kept all of us going was the expectation of what all this hard work would bring to the light of day.

Scene III: *Original panels. At the campsite the young Wakinyantawa, Kalsatakee, and Situakee sit in a circle. The evening meal is finished, and they are trying to converse.*

WAKINYANTAWA (EARTH WALKS): In the quiet of the evening, I shared food and talk with Kalsatakee and Situakee. It wasn't exactly talk, it was mostly sign language, drawings in the dirt and on pieces of rock using the red ocher paint, and frustrated gestures when we could not make each other understand.

By these means I found out the two of them came from the southland. They had traveled down a wide river, a river their village stood beside, until it emptied into the Father of Rivers. They paddled up this great, muddy current until they came to the markings, painted rocks, and tree signs that showed all travelers the way to the sacred grounds. From there they had walked overland to reach the quarries.

Kalsatakee and Situakee told me they came from a large village, a village that never moved and contained more people than a herd has buffalo. The village was enclosed by a high wooden wall, and all the lodges were built on top of high mounds of earth. If I understood correctly, the reason they did this was to get close to the sky spirits, as well as to show they were above creatures who crawled and swam. This I didn't understand, for all creatures were of Mother Earth, and one could not be above another.

I then explained I came from the grasslands to the west. That my village was much smaller than theirs, and that my people moved with the sun and the moon. I told them we were hunters of buffalo, and here I drew pictures of the way we hunted buffalo and told them of the making of pemmican.

In this manner time walked by. Days became weeks, and during these weeks other quarriers arrived. Some came from the south and talked of the great salt sea, others came from the west and told of high hills that sometimes rumbled and roared and glowed with fire, and yet others came from the north and told of sugar trees and food that grew on water. Soon the valley was filled with the music of many hammer-stones, music that causes the creator to smile.

It was a hot, muggy day, one with little wind, so the smell of a moving bison herd lay thick and heavy across the valley, when I uncovered the sacred stone. I remember my hands shaking as I cleared away the last of the hard, pink rock. Exposed below this was a thick, rich layer of blood red rock that changed, over the length of my arm, into a beautiful section of rainbow rock.

I could hardly contain my joy; I wanted to shout and run and dance. I felt light and young. The spirits smiled on me, and the work would now be pleasant and sweet, for this was the part of quarrying I loved. Working to remove the red rock, I could dream and imagine the kind of pipe each piece held within it, and how I would carve it to bring out what the spirits wanted the people to see.

I worked six long days, breaking and cutting the red stone into large pieces, then prying them out, watching them peel away from the underlying rock like skin off a cooked fish. I quarried the red stone first, then the rainbow rock. During this time I collected enough good pieces to make all the pipes our people would need, and enough again to trade. But it was not until the seventh day I levered out the piece I had asked the guardians to grant me. A rectangular piece, a hand and a half thick and over four long. The piece was the color of a summer sun seen through a thin veil of clouds, pale red with streaks of pink and white across it. Being as careful with it as a mother with a new baby, I took all afternoon to break it free.

When it finally came loose, I grasped it with both hands and held it high above my head so I could admire it in the light of the sun. The piece was magnificent, the perfect gift, a father could give no better to his son. I was truly blessed.

At that moment I heard the noise that will haunt me for the rest of my life: a rustling in the grass above my head followed by a long, low growl. Turning, still holding the rock high, I stared into the pale yellow eyes of a large coyote. Startled,

I worked six long days, breaking and cutting the red stone into large pieces, then prying the pieces out.

for it was rare to see one so near a working quarry, I stepped backwards. As I did, my foot slid between two pieces of rock. Losing my balance, I fell heavily to the ground. The piece of stone flew from my hands and crashed to the floor of the pit, splitting open like a blueberry stepped on by a bison.

Lying there, I stared in disbelief at the broken pieces. My gift, given me by the spirits, was ruined. There wasn't a piece left big enough for a regular pipe, let alone the pipe I was going to make. I had carelessly lost what the spirits had granted, and shame burned through my body.

The coyote yipped twice, long and high. I looked up and saw the creature had not moved. It sat at the edge of the pit staring at me, its long, pointed head tilted to one side, lips drawn back, exposing sharp, white teeth, all of which gave the beast a kind of lopsided grin. Shaking its head, as if it disapproved greatly of me, the coyote stood a moment more. Then, turning its tail toward me, it gave itself a good shake and vanished into the tall grass.

With that last shake I understood. This was no ordinary coyote. It was the nosy thief, the trickster who took the form of a coyote, the spirit god we call Old Man Coyote. He had come to steal the gift the spirits had given me, and in a way he had, for now it was no use to anyone.

I clearly remember reaching out and picking up one of the broken pieces. Gently rubbing it back and forth between my fingers, I wept for the pipe that wouldn't be, for the gift my son would never have.

I sat in that quarry and cried, cursing myself and my stupidity until the sun was on the edge of the land and long shadows stretched across the pit. The coming darkness forced me out of my misery and back to camp.

This was, as fate would have it, Kalsatakee and Situakee's last night at the quarries. They had all the stone they could carry. Also, they said the growing season would be getting on and they needed to be back to help in the harvest celebrations.

I hated to spoil their last night, but I felt compelled to tell someone the story of my disaster and the trickery of Old Man Coyote.

They agreed it was not my fault. They assured me I was a victim of that thieving trickster and urged me to have faith the spirits would deal with the evil creature. They also thought the guardians of the quarries would somehow find a way to return that which had been so unjustly taken. Because of this belief, they made me promise to go at first light and talk to the spirit guardians that dwelled beneath the sacred rocks, to ask their forgiveness.

This I agreed to do.

(The lights go down, and in the dim light, one lodge is removed from the stage, while the other is opened so the audience can see inside. A large object is also placed by the three maidens. When the lights come back up, a tape recording of blowing wind begins to play in the background.)

The next morning, I woke to bright sunshine and the blowing wind. The wind was a typical grassland wind, one that is born far to the north and west and, with nothing to slow it down, builds up force as it races south across the flatness. As you children know, a grassland wind is also a crying wind, a wind lost in the vastness of this green land, so it is constantly moaning and screeching, sounding to me like the cries of a wounded hawk. On this morning the wind was not lost; it had found my lodge to bother, and was now busy trying to tear the buffalo skins from the wooden frame.

Blinking sleep away, I decided, wind or no wind, to fulfill my promise to Kalsatakee and Situakee. I would go and ask the spirit guardians for forgiveness, and beg them to guide me to another worthy piece of red stone.

When I crawled outside, my spirits sank. I was alone. The other quarriers had left for the day's work, and Kalsatakee and Situakee had departed in the time before the coming of the light. Their camp had vanished, and the only reminders that it had ever been were the earthen mound, the fire pit, and the refuse hole.

Bending into the wind, my ears full of its ringing screech, I set off along the same path I had taken when I first arrived.

I was alone.

The path wound its way through the high grass, following the edge of the sacred valley. I moved slowly, fighting the wind and my own thoughts, for I was having a hard time deciding what I would tell the guardians.

It was the flapping of the deerskin that caught my attention and brought me to a halt. There, in front of me, sitting in the middle of the path, some five paces from the sacred rocks, was a small earthen mound. Lying on top of the mound was a decorated deerskin bag, the front of which had been blown open and was flapping about in the breeze. On top of the deerskin sat one of the clay drinking vessels Kalsatakee and Situakee used. The location of these objects, and their distance from the sacred rocks, told me these were not gifts for the guardians, but were meant for me. But by the Great Spirit, why would they do such a thing? And here? What could be so special that they wanted me to find it by the sacred rocks?

Putting the drinking cup on the ground with a heavy stone inside to hold it in place, I saw the deerskin had been recently painted with red ocher. There were two figures on the bag. One was a man, arms outstretched, offering a red pipe to a second, smaller man. Above both of them was a drawing of the sacred rock, only the rock smiled as it ate a coyote.

Here, see for yourselves. This is the very skin. The one Kalsatakee and Situakee left. I have kept it all these seasons. Careful now, for like me it is old and wrinkled and must be treated gently.

When I saw the drawings, my heart began to race like a scared fox. I realized their significance and knew why the two of them had been so insistent on my coming here. Hands trembling, I opened the pouch. My breath stuck in my chest, and like the great stone wall I stood frozen and still.

Not possible, I thought. I must be asleep and dreaming. Slowly, afraid it might vanish, I ran my finger over the smooth, rocky surface. Solid and cold—real stone, not magic and dreams. Before me lay a thick piece of the red stone, a piece shot through with pink and cream and dark red spots, some of which had long, white tails trailing behind them. This was a magnificent piece of rock, and must have been one of Kalsatakee and Situakee's finest, yet they had left it as a gift, a great gift for me to carve, and for my son to have. What a pipe I would make from this, a pipe worthy of the spirits, one carved with the deepest part of my soul.

As I caressed the stone, thinking of Kalsatakee and Situakee and dreaming of the pipe I would make, I smiled and looked up. I looked straight at the sacred rocks, and I will take a vow that just for the beat of a hummingbird wing, I caught the stones smiling back at me.

CHILD ONE: For the guardians and your son— did you carve a beautiful pipe?

WAKINYANTAWA (EARTH WALKS): A magnificent pipe. My son has it still. If you go along to his lodge and ask ever so sweetly, he may show it to you and tell you of the good fortune and power it has brought him. For it is a pipe carved from the soul and blessed by the spirits.

CHILD TWO: Did you ever go back to the quarries?

WAKINYANTAWA (EARTH WALKS): Bah. I did, and a waste of time it was too.

ALL CHILDREN: *(in unison)* Why?

WAKINYANTAWA (EARTH WALKS): When I returned home and told everyone the story of Kalsatakee and Situakee, Old Man Coyote, and the rainbow rock, old Wisherenedr, the one who could see into tomorrow, said I should not be so eager to lay blame on Old Man Coyote. I asked her just what she meant. She told me there may have been more to what the coyote did than mere trickery.

"And what could that be?" I demanded.

"Omens. Messages," she solemnly replied. "If you had kept your eyes and mind open you might have seen Old Man

Coyote was trying to tell you something. Think, is it not possible he was warning you, saying never again would a rock such as your rainbow rock be quarried on sacred ground?"

I thought she was crazy. She had never been to the quarries. She didn't know the rock as I did. I ignored her and her warning. I thought, next season, when I brought home a pile of rainbow rock, I would have my satisfaction—as well as her apology.

So I went. The weather was lousy and so was the red rock. Two more seasons I went, and there was no rainbow rock to be taken. The old woman had seen, she knew. So remember, all of you, listen to your elders, pay attention to what they say and know. It will save you a lot of pain and time and worthless effort.

(Earth Walks stands and walks to the edge of the stage and steps out of character.)

EARTH WALKS: Speaking of time, there was a time when the pipestone quarries could have been considered the United Nations of prehistoric America, a place where Native Americans from all parts of the continent, speaking many different languages, each with their own customs and rituals, could gather and live together in peace and mutual respect as they worked toward a common goal: to respectfully quarry the red rock, and from it make sacred pipes for their people.

Today this comparison is even more true—but unfortunately that is not necessarily progress, or a good thing. Pipestone, just like the United Nations, has turned into a place of contention and conflict. Many of today's Native Americans, it seems, cannot agree on the definition of 'sacredness' when it comes to the display and sale of the pipes carved from the red rock.

"Some Native Americans claim the pipe carvers are selling sacred objects to the public and should be prohibited from doing so. They have even gone so far as to hold protests at the carvers' own workshops and at places where the carvers' work is displayed.

Mostly, however, the carvers sell their work to medicine people so the pipes can be used in ceremonies. Doing this simply replaces the trade that was done in olden times. For early Native Americans, as for the modern carvers, a pipe becomes sacred only when used in ceremony; otherwise it is but a beautiful work of art. The protestors also consider the scrap rock from which the pipes are carved to be sacred, and thus frown on the making and selling of earrings, pendants, and other small objects carved from this scrap. However, in using this scrap rock the carvers are following the ancient tradition of not wasting anything they use, much like the Native Americans who found a use for all parts of the buffalo. Sale of such small objects allows the carvers to earn a modest living in today's society, a living that lets them continue the ancient tradition of quarrying and pipemaking.

This divisiveness, which is what the creator was trying to avoid when he called the people back to the quarries and fashioned the first pipe of peace, reminds me of something Sam Rayburn once said: "A jackass can kick a barn down, but it takes a carpenter to build one."

I believe the same thing is true of pipestone, traditions, and the United Nations!

CHAPTER EIGHT

THE TALKING SKY

PART ONE

FISHER AND FRIENDS

I hate deerflies. In fact, I think most geologists do. But then, when you work out of doors in remote places, especially where it is hot and dry, who wouldn't? The incessant buzz and hum as they circle round and round your head, small sharks waiting for a feeding frenzy. Clouds of them swarming over you like ants over a pot of sugar every time you stop to examine an outcrop. Insect spray, even the stuff that takes the varnish off pencils and dissolves plastic, does no good. I've drenched the back of my hands with such poison, and as soon as I've applied it, a deerfly will land in the thickest pool and with great gusto dig right in.

It is said that for every action there is an equal and opposite reaction. The reaction of a geologist who works in places where deerflies outnumber oxygen atoms is to devise unique ways to kill the horrid biters.

Simple things to please a simple mind, I thought as I killed another deerfly. But for me it seemed "simple" was habit forming, for it was another simple thing, like a question, that ended with me standing in the middle of nowhere, repairing a birchbark lodge so I could sleep on a pine bough floor. All for comfort and to be free of biting insects!

It all started simply enough as well—all I did was read a book on archeoastronomy. This was a subject I knew nothing about but found fascinating. In an attempt to find out if there were similar books on the star lore and astronomical knowledge of the Ojibwe and Dakota I called Earth Walks. Before I could say Big Dipper, I found myself in the middle of the north woods.

Earth Walks, unbeknownst to me, was an authority on the subject. He happily agreed to share whatever Native American sky knowledge he possessed, but only if it was done properly. Now why wasn't I surprised!

To properly learn about the Ojibwe sky, according to Earth Walks, we needed a place far from city lights, one with an unrestricted view of the heavens. And he knew the perfect place.

That place was the campsite on what he said was his family's off-reservation trust land, a small clearing on the edge of a long, rocky lake, surrounded on three sides by cathedrallike trees. The trees were part of a magnificent old-growth forest of red and white pine, paper birch, balsam fir, and quaking aspen, with red oak and maple thrown in for good measure. Some of the white pines were so large Earth Walks and I together could not get our arms around them.

The camp turned out to be a birchbark lodge, which we spent the good part of a day making water- and fly-proof as well as comfortable, and a sweat lodge, which we spent the rest of that day constructing.

Behind the camp the forest rose up to cover a high ridge of granite. At the top of the ridge, directly behind the camp, was a bare granite knob—Earth Walks' perfect place.

This large outcrop provided a magnificent view of the heavens, horizon to horizon, so that someone sitting at its center would feel as if they were inside a planetarium. The throne of a sky lord, by Earth Walks' reckoning. This natural observatory was a twenty- to twenty-five-minute hike from the camp, along a well-maintained trail.

At dusk, on the second evening in the camp, I found myself sitting in the middle of the sky lord's throne. Earth Walks had built a small fire, and we drank camp coffee as we waited for darkness to enfold us.

"Funny thing about people and fires," Earth Walks was saying.

"Such as?"

"An Indian, like me, builds a small fire and moves close to it; a white man, like you, builds a big fire and runs away from it."

"Meaning I should move closer?"

"Up to you," he said, with a shrug of his shoulders. "But it is getting chilly."

"This land been in your family long?" I asked, taking Earth Walks' Indian advice.

"Couple hundred years at least. My grandfather was the last to do real trapping on it. He came up here all through the thirties and forties. Once or twice things got so tough he would have liked to have sold it. But he couldn't, for it was what is called an outside allotment, kind of like a reservation off the reservation. As it turned out, it was a good thing he couldn't."

"I'll second that," I said. "It's beautiful."

"You know," he said, filling our cups from the blackened pot that sat at the edge of the fire, "my interest in the night sky and Native American sky stories began right here when I was twelve years old. My dad, even though he didn't trap or hunt much, brought me out here in early spring to teach me about the Ojibwe's life on the land, as well as to become aware of the wonder and importance of the stars. He told me my grandfather had done the same for him, and he hoped I would continue this with my own children."

"Sounds like the start of a great family tradition."

"Not so much tradition," Earth Walks replied, "more of passing on a way of life. Without TVs, VCRs, books and computers, Native Americans spent a lot of time star gazing. In fact much of their everyday lives—from rituals to hunting, gathering, or farming—was directly influenced by celestial patterns of space and time; for them it was not Fox and a three-star movie, but Mother Earth and Father Sky, creation, the land, and the cosmos."

"For me," I commented, "it sounds more like Gaia, at home with *The Hitchhiker's Guide to the Galaxy.*"

Earth Walks laughed loudly. "Pretty fair for a white man," he said.

"Speaking of patterns in space and time," I continued, "you were going to tell me something about Native American rock paintings."

"So I was and so I will," he replied. "First you have to know that I believe some of the rock paintings, or pictographs, represent constellations, pictures Native Americans saw in the sky. The Ojibwe, for one, have many folk tales where the main characters end up inhabiting the sky world—a world visible to any who bother to lift their eyes heavenward."

"Second, not everyone would agree with me. You see the pictographs found around here, as well as hundreds more across North America, have been analyzed, photographed, drawn, stared at, and puzzled over in more ways than there are to cook buffalo meat. Yet, with all these impressive investigations, their purpose and meaning, as far as archeologists and historians are concerned, remains largely an unsolved mystery."

"And the artists, whoever they were, always painted on rock outcrops?" I asked.

"Or glacial boulders," he said, stirring more sugar into his coffee. "The paintings are usually found on vertical or sloping rock faces along lakes and rivers that were part of great Indian water highways, trade routes that were interconnected and well traveled, or routes that led from winter bush camps to spring sugarbush towns or summer villages. Some are also on ridges, open and high, good vantage points and easy walking. Rock drawings in Alberta, for instance, are common along old buffalo migration routes."

Unexpectedly he laughed. "It has been suggested that many were painted from canoes. Imagine! Bobbing up and down, trying to stay in one place, stable and steady, as you paint, no wonder many of the figures are said to lack a lot of detail! It seems more reasonable that many were painted during late fall, winter and spring, when the lakes and rivers were frozen, and travel took place by snowshoe and sled."

"With so many pictographs, I suppose there is a great variety of subjects painted?"

"More than the patterns in a quilt at a quilting bee," he answered, shaking his head. "You can see abstract symbols; supernatural beings, such as human shapes with horns or birds heads; many different animals, including moose, fishers, panther, elk, deer, bear, buffalo, and so on; as well as humans."

"I did read," I commented, putting the last piece of wood in the fire, "that many of them are thought to be a thousand or more years old. Is that right?"

"They are difficult to date accurately," he responded. "However, based on subject matter and tools shown, many are believed to be in the range of 500 to 1,500 years old. Some, however, show muskets and sailing ships and were done in fur trade times.

"The pictographs have lasted so long," Earth Walks continued, "because of the pigment and binder used for the paint. The pigment was a powdered, or finely ground, form of the mineral hematite."

"Iron oxide," I offered. "That's why most of the pictographs are reddish in color?"

You can see abstract symbols, supernatural beings, animals . . . as well as humans.

Earth Walks nodded. "Actually brick red to dull brown," he said. "For painting, the powdered hematite was mixed with a binder of either fish oil, sea-gull eggs, or sap from certain common plants, such as milkweed, to give it a permanence that would make many modern paints run with envy. The Indian paint is difficult to scrape off, and in many instances commercial paint that vandals use to cover the pictographs has already worn thin or largely disappeared, leaving the Indian pictures to shine through. A good example occurs at Agawa in Lake Superior Provincial Park, Ontario. Black paint, used in 1937 to paint over a large pictograph, is wearing away like a street in Duluth, while the Indians' red ocher still looks like last year's highway. The pictographs at Agawa are at least two hundred years old, and over this time they have been severely weather-beaten by waves, wind, and the shore ice of Lake Superior; but unlike the paint on many houses, they remain fresh for everyone to admire."

"Astronomers, artists, and inventors of everlasting paint," I offered. "Pretty impressive people. So why did they paint the rocks in the first place?"

"Actually," he said, "no one knows for sure. But answers and suggestions are as numerous as bears at a garbage dump. Some of the possible reasons include—" and he held up his hand so he could tick the following points off on his fingers:

(1) Ceremonial: a plea to the spirits to give the painters the good things in life, such as health, fertility, rain, and plenty of game. They may also have played an important role in coming-of-age and healing ceremonies. Pictures of animals the Indians hunted may have been drawn on the rocks to appease the animals' spirits and make certain they would return to again be harvested.

(2) Dreams: dreams or visions were important to many Indian tribes; some of the paintings could have been done during 'vision quest' ceremonies or rites.

(3) Mnemonic reasons, as memory aids: since Indians had no written language, the pictographs may be records of legends, tallies, seasonal changes, and so on. Selwyn Dewdney, author of *Indian Rock Paintings of the Great Lakes*,

relates them to Ojibwe birchbark scroll paintings, which were used to keep track of such things as the order of ceremonial songs and the steps in certain initiation ceremonies.

(4) Recording of important events: for instance, there is good evidence that the Anasazi people of the Southwest U.S. recorded in pictographs the supernova that occurred in the Crab Nebula in A.D. 1054 The pictographs may also be warning signs, telling of something bad or evil that occurred at a particular place. On Rainy Lake, in northwestern Ontario, for example, a series of pictographs possibly tell of the overturning of a canoe, the drowning of two men, and the way one man managed to save himself.

(5) Clan symbols: these are the totem animal or object that is considered to be the guardian symbol of a particular clan. Membership in a clan is usually inherited by birth, and an individual would be identified as a member of the bear, elk, eagle, wolf, or other animal or plant clan. Clan symbols painted on, or carved into, rocks are not uncommon in the southwest U.S., and at Pipestone may have been used to indicate who was quarrying where.

(6) Fun: doodling to pass away an idle afternoon or a cold, clear winter day. Some could have been done by children as a means of keeping them busy, a learning tool, or tribal activity.

(7) And finally, as I believe, astronomical: Native Americans had a great interest in the stars and other heavily bodies, from comets and shooting stars to planets, the sun, and the moon. The native people, since they spent much time watching the heavens, most likely made up mythical creatures in the sky, or visualized actual animals they hunted, feared, or lived with. These star creatures, over the ages, would then become symbols of important changes in the yearly cycle of the tribe, animal migrations, floods, plants ready to be harvested, maple sugar time, time to end the winter bush camp, and so on. It has been

suggested by Thor and Julie Conway that many of the pictographic images represented constellations; as possible proof, they point to Ojibwe folktales in which the characters end up in the sky.

"Overall I think my grandfather summed it up nicely in a poem, one my father remembered and passed on to me. According to grandfather, when you went out and looked at the rock pictures you needed to remember that the Native American artists who painted them were also:

> Dream catchers,
> Star gazers,
> Earth watchers,
> Myth makers,
> As well as painters of the rocks.
> They came
> By boat,
> Along the trade routes,
> By snowshoe
> Along the shores,
> With earth dreams
> To paint on mornings cliff face,
> And star scenes
> To draw on evenings shore,
> They were the painters of the rocks."

By the time Earth Walks finished his poem the fire had burned low, and the sky had darkened enough to reveal more stars than the forest has trees.

"All right," Earth Walks said, rubbing his hands together the way ministers do before an important sermon. "Enough of rock mysteries. It's time to begin to change your perception of the heavens. First," and he pointed to a spot high in the sky, "there is what the Romans and Greeks call Ursa Major."

"The Big Dipper I know," I replied sarcastically. "I was once a Boy Scout. I didn't realize the Big Dipper was also an Ojibwe constellation."

"Bah," he grumbled, "be quiet and listen. Your Big Dipper is a small part of what the Ojibwe call Fisher, a fisher with an arrow in its tail. Come over beside me and I'll show him to you."

As I moved next to Earth Walks, he raised his arm and said, "The tail begins where the dipper's handle does. The four stars of the dipper are the fisher's body, and those stars there, to the east, are its head and feet, while those three stars right there represent the arrow in his tail. Can you see him?"

"About as much as I do any creature in the stars," I answered. "I need time, but you certainly have a powerful imagination."

"Not as good as the Greeks and Romans," he quickly replied. "In the early spring the Fisher swings upward from the northeast. Here, let me show you." And he pulled a star chart and a small pen light out of his pack. Turning it to early spring, he showed me how the Fisher rose above the horizon.

"This was of great importance to the Ojibwe," he remarked, turning off the l ight. "It meant the birds of summer would not be far behind."

"Birds!" I exclaimed. "What do birds have to do with a sky fisher?"

"Everything," he answered. "Let me tell you."

And as I tried to connect star dots to make a fisher with an arrow in its tail, Earth Walks told the following story:

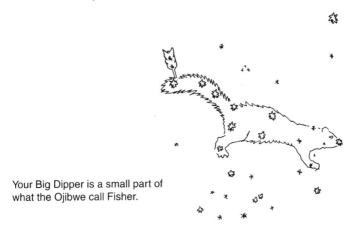

Your Big Dipper is a small part of
what the Ojibwe call Fisher.

Once Fisher and his friends were out on a hunt. The hunt lasted weeks and weeks. The hunting was very difficult because the snow and cold would not leave them alone. Fisher's friend, Bear, began to worry.

"Winter has lasted too long," he told Fisher. "If spring does not come soon we will all starve. The moose and caribou will have nothing to eat. The beaver will have no lily roots or fresh aspen bark. Something has happened in the sky world to stop the seasons from turning as they should."

"Let us send Wolverine up to the sky world to find out what the matter is," Fisher said.

They sent for Wolverine. He agreed to go, and ascended to the sky world by way of a great pine tree. He was gone for many days. Finally he returned.

"A great ogre beyond the edge of the sky has captured all the birds," Wolverine reported. "He has imprisoned them in great birchbark makoks. That is why winter will never end."

"Who is this ogre?" asked Fisher.

"He is bigger and more cruel than any being here in this world." Wolverine said. "Worse, he has his brothers with him to guard the birds."

"We must kill him and free the birds," said Fisher.

Having said this, he strapped on his quiver and his knife, picked up his bow, and set out. He came to the great pine tree and climbed it. From the top of the tree it was but a short leap to the opening in the sky.

Once through the opening, Fisher found himself in a wonderful world. It was warm, flowers were everywhere, and the air was alive with the buzzing of bees. Moving across the land, Fisher soon came to the ogre's encampment.

The two guardians turned to face him. Realizing quickness was his only chance Fisher dashed between their legs. He ran as fast as he could to the huge baskets and stove them open. Out poured the birds: flickers, jays, robins, chickadees, ducks, geese, and swans; up they spiraled in a great black cloud that darkened the entire sky. Then, in a tornado of wings, they plunged down through the hole in the sky and entered the world below.

The great ogre shouted in rage. He and his brothers ran toward the brave little fisher. Once again Fisher used his speed and quickness. He

Fisher and friends.

dashed between their legs and raced to the hole in the sky. Without hesitating, he threw himself through it. Far below he could see the earth. And before his eyes it was changing from white to green. Down he fell, the ogres' arrows whizzing all around him.

Fisher was lucky. None of the arrows found its mark, and he landed on soft, mossy ground. He knew the ogres would not be far behind him, so he had to make his escape fast. He ran this way, he ran that way, he dodged terrible flights of arrows. But try as he might, he couldn't lose the ogre or his brothers. In fact, they were getting closer.

In desperation he raced back to the great pine tree, thinking he could fool them by climbing into the sky world, then doubling back to earth. Quickly he climbed the tree, but it was not fast enough. The ogres saw him, and a great volley of arrows whizzed by, missing him by inches. At the top of the pine tree Fisher leapt to the northward. Here one of the arrows found its mark and pinned him to the sky. Around and around he turned, and there he is to this very day.

With the freeing of the birds the ogres lost their power over the earth. So they left by way of the great pine tree, back through the hole in the sky to their world. They have never bothered the inhabitants of the earth again.

"Sad story," I commented. "You Indians certainly have nothing on the old Greeks."

"Rot!" he exclaimed with disgust. "It's not sad at all. In fact it's a story of great bravery and courage and its rewards. Fisher didn't die. Each year he circles the heavens. In winter he crawls along the northern horizon, but in the spring, he gains power and ascends high overhead. That is when the birds come, great flocks of singing birds, water birds, fishing birds, and birds of the prairies and forest. Then, slowly, Fisher swings back to the west toward the horizon, and it is time for the birds to leave. Thus Fisher, in gaining power, is the herald for the coming of the birds, and when he loses that power, he is the messenger that tells them it is time to go. The birds of our land, they come back every spring when the brave little fisher rises high in the night sky. Because of Fisher, they know when spring is in the air, and so do the Ojibwe."

"Pretty neat, I'll grant you that," I told him. Pointing to another figure on his sky chart, I asked, "What about this one, the one labeled Panther. Does he have as brave and courageous a story as well?"

"Great Panther," Earth Walks corrected. "And he is another matter all together." To find him you first have to find Regulus in the constellation Leo; it's the bright star right there. The three stars above it form a sicklelike shape, which represents the

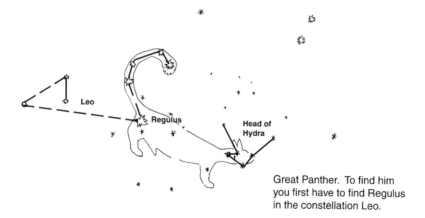

Great Panther. To find him you first have to find Regulus in the constellation Leo.

panther's tail, while Regulus is the start of its hind leg. Those five stars way over there, in the east, the ones that form the head of Hydra, represent the panther's head. So by using those stars we can now draw the panther in the sky."

And Earth Walks proceeded to do just that while I followed his moving arm.

"And the story attached to the tail of this panther," I asked, while attempting to duplicate Earth Walks' celestial drawing.

"In order to understand the panther's story," he said, "you have to know that early spring is a dangerous time to be in the north woods. This is the time the ice goes out. The snow turns wet and soggy, and you can hardly walk through it. Your snowshoes become so heavy they feel like you have iron weights on your feet. The rise of the Great Panther in the spring sky signals danger, possible death, and the time for great floods.

"The story that goes with this constellation has many variations, but the one my grandfather and father told goes like this:

Nanaboujou was out hunting with his cousins, the wolves. Seeing them coming, Nanaboujou's friend, the woodpecker, called to him from a high tree: 'You must watch out, Nanaboujou, a terrible monster is laying waste to the country and it wants to destroy you.'

Nanaboujou knew the monster was Curly Tail, the Great Panther, who had left his home on the bottom of Gichi Gami and was out looking for him.

"Take a feather from my crest," the woodpecker called to Nanaboujou. "Put it on your arrow. This is the only way you can defeat the mighty Curly Tail."

Nanaboujou did as the woodpecker suggested.

Then Curly Tail emerged from the pine forest. He snarled at Nanaboujou, and his screams echoed across the great lake. Curly Tail changed his form to a long, golden-scaled monster with horns. The monster rushed at Nanaboujou. When it got close enough for a killing leap, it changed back into the enormous cat. Nanaboujou stood his ground. As the cat started his death leap, Nanaboujou drew back his bow and loosed the arrow. It flew straight into Curly Tail's side, and the cat's roar of pain caused the wind to stop. Curly Tail rushed back to the great lake. Plunging into the water, he disappeared beneath the waves.

Nanaboujou knew Curly Tail was not mortally wounded, and he feared greatly that the cat would cure himself and return to take his revenge on the world. He knew he had to finish the kill. Approaching the shore of the great lake, Nanaboujou spied the evil frog people.

"Nanaboujou!" they called out. "Curly Tail has offered the riches of the world to anyone who can cure him."

"Lend me a frog skin," Nanaboujou said. "I will cure him myself."

Covering himself with the frog skin, Nanaboujou dove to the bottom of Gichi Gami. He passed by the panther manitous who stood guard outside of Curly Tail's lair. Slowly he approached the great cat, who hissed and snarled in pain.

"Have you a cure for the great Curly Tail?" the panther cried, gesturing to the arrow in his side.

"Of course," Nanaboujou replied. But instead of curing him, Nanaboujou leapt toward the panther and rammed the magic arrow in further.

Fighting his way to the surface, Nanaboujou was certain he had finished Curly Tail.

As he climbed out of the water, he noticed it was raining. A steady downpour that went on and on until the entire world was flooded. Nanaboujou only had time to launch his great canoe and assemble all the animals aboard. Nanaboujou now realized he had not killed Curly Tail; the panther was alive and taking his revenge on Nanaboujou by destroying his world.

Many days and nights they drifted across the watery world. There was no land to be seen, and Nanaboujou realized he would have to make a new world. But to do so he needed some earth to start with, and he had none.

He asked many of the water animals to dive down to the bottom of the sea and bring him back some earth. The beaver, the otter, the loon all tried, and all failed. Finally, the humble muskrat offered to try. The other animals laughed; hadn't they, the mighty swimmers, all tried and failed? But the brave little muskrat dove out of sight and was gone for many days. Finally his poor body was found floating on the surface, and there, clutched in his paws, was the precious earth. Nanaboujou twirled the earth in his hand, and slowly it grew. It grew so large it became an island. But it didn't stop there; bigger and bigger it got until it was finally the whole, dry earth. The animals leapt ashore, and so it was that the new earth was born from the old.

Curly Tail still lives at the bottom of Gichi Gami. His breath is pestilence, and with his mighty tail he can founder the biggest and finest canoes. For people to cross the water safely, Curly Tail must be given gifts and offerings. Only then will he spare travelers. And even though the Ojibwe are the finest canoemen in the world, drowning is the most feared of deaths.

In the spring the lakes in the north country, along with their creeks and tributaries, are known for rapid and violent rises in water levels, as well as for floods. For a winter hunter and his family, trying to cross this country after breakup would be next to impossible. When the panther reaches its ascendancy in the spring sky, it tells the people to get out of their winter camps, for the panther brings with it the promise of raging streams and spring floods.

"Impressive," I said. "Not only do your Ojibwe constellations have stories to go with them, but the stories actually reflect something important that is happening on the surface of the planet. Knowing the constellations, being able to read the stars, could well mean the difference between life or death on the land."

"As with geology," he said, "knowledge of astronomy was also necessary in order to survive on the land. Anyway, when the panther rose in the spring sky the Ojibwe left their winter camps and came together at the sugarbush. The sugarbush was a festive time. Remember, these people lived in isolation during the winter, scattered across the snowbound land. The sugarbush was the time and place they were reunited with relatives and friends. So it was a time of joy. But it was also a sad time. For there were those who did not make it through the winter. For those people a feast for the dead was held, and tobacco was offered for the ones who had gone."

"I assume the dead are also remembered in the sky," I commented.

"In life or death," Earth Walks answered, "the people of the Pays d'en Haut have a place on the earth and in the sky. When an Ojibwe dies the soul does not depart immediately; the soul lingers to say good-bye to all of his or her friends. Tiny spirit houses are built so the soul has a place to live. The soul only needs this for two weeks. Then the little houses are allowed to return to Mother Earth, and soon the brush, weeds, and trees take over the disturbed ground. But the Ojibwe always remember where their loved ones are buried, and they give a gift of tobacco in remembrance.

"When the soul departs, it travels a path to the west and faces many obstacles on its journey. First, it must eat of the great heartberry that pulses in its path, and then the soul must cross a deep ravine on a quaking log. Finally the soul comes to the great, shining river, a river whose reflection you can see in the night sky, and which you whites refer to as the Milky Way. The soul paddles this river to the great land beyond, a land where prairies and woods are filled with game. Here the soul greets all the friends and relatives accumulated since the very beginning of time. All is feasting and dancing; when we here on earth look up and see the northern lights playing across the sky, it is our relatives dancing in the sky."

"I like that," I said. "It's a nice explanation for the Milky Way and the northern lights."

"Then you should also know that the great shining river is the Ojibwe's equivalent of Christianity's heaven and hell."

Looking at the white arc, stretched like a rubber band across the heavens, I asked him how that could be.

"Look to the light shimmering off the great river, your Milky Way," he said. "See, down towards the south how the river forks. One fork leads to the beautiful place I have just described—heaven. See the other fork, the one that disappears into nothingness—your hell."

"I see it. I surely do!" I cried. "Save me, Reverend, save me, for I have sinned. But now I have seen the light, I truly have!"

"Pfui," he grumbled, "for you it's way too late."

Laughing, I asked Earth Walks about some of the other animals he had named on his sky chart—a moose, a canoeman, and three canoes, among others.

"Ah, those," he sighed and smiled. "They will have to wait."

"What for?" I asked.

"Spring," he answered. "These are winter and spring constellations. When we go to Hegman Lake next spring I will introduce you to them, and with their help, plus that of the great panther, I will show you the connection between pictographs and the Ojibwe sky."

"You've got to be kidding!" I exclaimed, thinking the chances of me being found at Hegman Lake in the spring were about as good as sitting down to a Newfie squid burger—zero or less.

"The hour is getting on, "Earth Walks said, stifling a yawn. "But before we head back, there are two more constellations I want to show you. See the circle of stars right over there that has a small break in it?"

I looked to where he pointed and nodded.

"Those are what the Ojibwe call the sweat lodge, and the break in it is where the poles have been torn out. Over there, right next to the sweat lodge, is what is called the exhausted bather. Good, isn't it? Now that we've found the sweat lodge in the sky, we

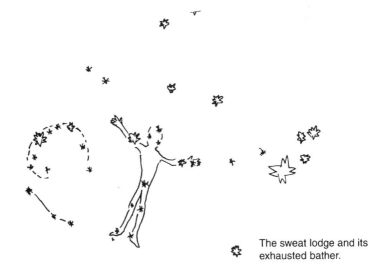

The sweat lodge and its exhausted bather.

can head back to camp and find the one we built. Let's see if we too can become exhausted bathers."

"I for one won't have to try too hard," I said as I shouldered my backpack and turned on my flashlight for the hike down the granite ridge.

PART TWO

THE IMMENSITY OF IT ALL

Early spring, Hegman Lake stiff with ice, and me covered by so many layers of clothes I could have been related to the Michelin Man. At that particular moment in my existence, the main thing I was grateful for was that Earth Walks was not a seller of squid burgers!

Sipping the scalding liquid he referred to as coffee, I wiggled further into the portable lean-to trying to escape a biting north wind. It was dusk, a Hegman Lake dusk that was much different from the ones I had shared with Earth Walks last summer on the sky lord's throne. This was a quiet dusk; no creaking ice,

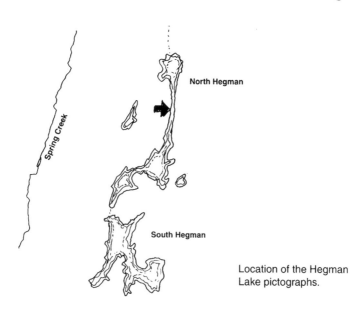

North Hegman

Spring Creek

South Hegman

Location of the Hegman
Lake pictographs.

N

squeaking trees, howling wolves, crashing moose, singing birds, crunching snowshoes, or howling snowmobiles—it was as though we were alone on a silent planet. Though the calendar said it was early spring, this was surely a winter quiet, one that fit perfectly with the immensity of the deep blue sky, clear as Irish crystal but for the darkness growing around its edges.

We had been at Hegman Lake long enough to have a close look at what Earth Walks called the most photogenic of all the Canadian Shield pictographs. They gave me goosebumps, a sense of long ago, important comings and goings, changing seasons and people, all taking place under the outstretched arms of an imposing male figure. The figure, according to Earth Walks, was a good caricature of the ideal canoeman: deep-chested; powerful arms; short, muscular legs, perfect for kneeling in a canoe. In the rock painting his arms are stretched wide, possibly, so Earth Walks thought, to embrace everything the Great Spirit had put on earth.

Beneath the canoeman is a catlike figure, one Earth Walks believes to be a panther, though he did say the identity of this animal has always been debated. The panther, as I already knew, had been the focus of Ojibwe tradition for hundreds of years and figured into several creation myths where Nanaboujou and "Curly Tail" cause the great flood.

Hegman Lake pictographs—
main group.

To the right of the panther is a large moose, and directly above the male figure are three canoes, the same three figures I had seen last summer on Earth Walks' sky chart. High above these was a cross, and beneath the entire group of figures was drawn a straight line. The figures were clustered together on a narrow strip of granite, and to their left were six short horizontal marks and three crosses in a vertical line. Just above the moose's head was a short, tapering stroke of paint. The paintings faced due east and were easily visible from the lake.

Earth Walks also showed me where, in each figure, the artist had erased an area designating the heart. He told me this practice was common among people who lived on the northern plains and in the southwest U.S., and is called the "heart line." It also appears in the art of the Canadian Ojibwe artist Norval Morriesseau. To Earth Walks, an artistic convention like the heart line, one that has apparently endured for centuries, gives strong support for comparing the Canadian Shield rock paintings with Ojibwe traditions and myths.

After swallowing a bit more of the hot liquid, I felt warm enough to unzip my parka. Taking my gloves off, I wiggled my fingers and asked Earth Walks how he first got the idea of connecting pictographs to Ojibwe constellations.

Pushing the hood of his parka off his head, he drank his coffee before answering. When he put the aluminum cup down he said, "I had been playing with the relationship for a long time. My grandfather and father sort of infused me with it. But I hadn't really done much, not taken it too seriously. Then I read about Ann Sofaer. She is an artist and is credited with the rediscovery of the Anasazi 'Sun Dagger' site, at Chaco Canyon in New Mexico. This site marks the summer solstice with a shaft of light shining through upright sandstone slabs to illuminate a petroglyph, which marks the solstice for yearly reference. This find proved these ancient people were sophisticated astronomers who measured time and kept exact calendars. If the Anasazi did it, I said to myself, why not the Ojibwe?

"I then read about the Bighorn Medicine Wheel, with its twenty-eight spokes, which sits at an altitude of 10,000 feet, close to the summit of Medicine Mountain in the Bighorn Mountains of Wyoming. It remained an ancient and mysterious rock construction until astronomer John Eddy demonstrated that its most likely use was to mark the summer solstice and the days of the lunar month, which many Native Americans had determined to be twenty-eight.

"These studies, along with the work of the Conways and my star lore background, eventually led to the Hegman pictographs.

"I came here with an idea and a sophisticated research apparatus," and he held up his star chart. "One of the expensive ones, you have to turn it to find the right constellation for the correct time, date, and latitude. And that's it—you pretty much know the rest."

"Now that it's dark enough," he said, fixing me with his owl-like look, "I'd like to see what you remember from our last star session. Come on out and show me the Great Panther."

That caught me off guard. There was a moment of fumbling to rearrange my thoughts and get up and out of the lean-to. Once in the crisp air, I managed to find the constellation of Leo, then Regulus and finally the sickle that formed the panther's tail. From that I pointed out the main stars of the panther, and, sweeping my arm across the heavens, I drew old Curly Tail.

"Not bad," Earth Walks cried. "Not bad at all. Now, take a look at this. It's a drawing of the Hegman pictographs. Without too much of a strain, don't you think you just drew a comparable figure to the cat shown here, the same one painted on the rocks above us?"

"I have to admit I did," I replied, "though I think mine is much nicer."

"When we get back," he scolded, "I'll get you some Indian paint and we'll see how nice your panther looks when you put it on a piece of granite!"

"Not fair," I shouted. "You know I can't draw a straight line without a ruler."

"Ruler!" he whooped. "I don't believe the painters of these rocks had any such things. Now, for the world's perfect canoeist. If you look at the star chart," which he illuminated with a pen

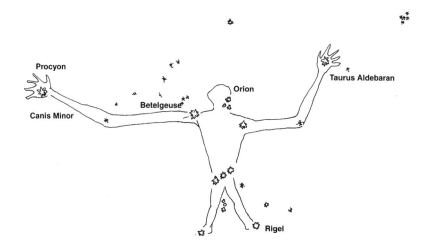

The Ojibwe Wintermaker, or Earth Walks' canoeman.

light, "I think you can see how it fits perfectly with the outline of Orion."

"The Wintermaker!" I exclaimed. "You're telling me your canoeman is the famous Wintermaker of the Ojibwe."

"Bingo!" he cried. "But instead of the upraised arms the Greeks saw, the great outstretched arms of the Wintermaker reach across the sky to the first-magnitude stars Procyon and Aldebaran—those two there and over there.

"Moving to the moose figure, I think the great square of Pegasus fits beautifully. And the faint constellation of Lacerta, which we can't see just yet, but which is here on the star chart, can be used for the moose's great antlers. What is even more fascinating is that in the great square of Pegasus is a star that can easily represent the moose's heart, right there," and he showed me on the star chart, then pointed upward to a faint star low in the western sky. "There also happens to be a 'bell' or 'beard' for the sky moose, although this is displaced somewhat in comparison to the one in the rock panting."

"So far so good," I said. "But your star figures, your constellations," and I waved my hand upward, "are much further apart than the ones in the rock painting. Isn't that a bit of a problem?"

"Good eye and a good question," he said. "The answer, I think, has to do with available space to paint, as well as the probability these figures were painted in the daytime from memory. And, as constellation diagrams, they certainly explain the stylization of the figures."

"All right. I'll buy that," I said. "But if these do represent constellations, then what is their significance? And why were they painted here?"

"Ah—the crux of the problem," he stated. "To tackle it from an astronomical point of view, I asked myself whether or not these figures, these possible Ojibwe constellations, could all be seen at the same time." Picking up the star chart he said, "Watch what happens around the middle of March."

He rotated the chart. I watched the panther rise in the east, and as it did so, the great moose moved toward the western horizon. And directly overhead, majestic and coldly powerful, was the figure of Orion, the Wintermaker.

"These are the positions of the constellations during the early evening in mid-March," he said. "That was about three weeks ago. At the spring equinox, however, you cannot see these constellations at this particular time; the setting sun obscures them. But about two weeks earlier, they come alive in the late winter sky; they appear together for the first time in anticipation of spring.

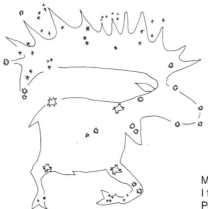

Moving to the moose figure, I think the great square of Pegasus fits perfectly.

"Think of it," he said, the excitement in his voice growing. "For the artist who drew these, this would be the time to move from the winter hunting camp to the maple sugaring village, and there to take part in what is called the spring thanksgiving ceremony. This was a time of rejoicing over winter's end, of remembering those who had passed on during the season of cold and snow, and the end of the isolation of the small bush camp. This sequence of events is nicely described by Basil Johnston. Here, allow me to read this to you."

Earth Walks took a small book out of the side pocket of his pack. Turning to a dogeared page, he read:

> In the spring with the first flowing of sap, there was thanksgiving. Such a public gathering coincided with the return of the Anishinabeg to community living after a winter of isolated living in family form. The principal theme for the thanks was gratitude for survival from the ordeal of winter.
>
> There was the smoking of the pipe; then offerings were given to Kitchie Mantou. After, there was a dance and chants, both embodying the theme of survival . . .
>
> We have endured
> The ordeal of winter
> The hunger
> The winds
> The pain of sickness
> And lived on.
> We grieve for those
> Grandparents
> Parents
> Children and
> Lovers
> Who have gone.
> Once again we shall
> See the snows melt
> Taste the flowing sap

Touch the budding seeds
Smell the whitening flowers
Know the renewal of life.

He closed the book, stuffed it into his pack, and said "After this I was convinced the three figures represented constellations. So I started looking for the other stars shown in the rock painting. The seven marks close to the Wintermaker's head could easily be the Pleiades. The star shape, or cross, drawn above the Wintermaker may be Capella, the brilliant blue star that shines directly overhead in late winter and early spring. In many Indian societies a cross is used to represent a bright star. Others believe this mark to be the sign of the Grand Medicine Society, which would mean this site is a sacred place. And this too is just fine, for it was the members of this society who were the tribe's chief stargazers!"

"What about the three canoes?"

"I couldn't find them," he commented, shaking his head. "Not in any star group, even though I searched the winter skies, nothing to make canoes out of in Cygnus, Perseus, Aries, Lyra, or any of the others. The only canoe was the one my grandfather and father saw in the W of Cassiopeia, but that was it. However, in thinking back to traditional Ojibwe stories and traditions, I realized the canoes might not represent star groups! In Ojibwe tradition the Milky Way represents . . . "

"The river of souls," I volunteered. "Your shining path to the afterlife."

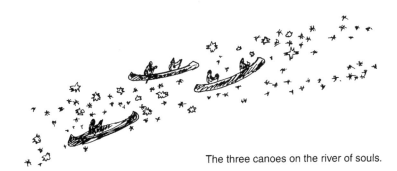

The three canoes on the river of souls.

"In the rock paintings," he smoothly continued, "the three canoes may represent the spirits of the deceased, traveling along this river of souls in stellar canoes, all the way to the afterlife. In early spring, as the three Hegman Lake constellations blaze in the sky, the great shimmering river is clearly and easily seen passing just to the left of your Orion."

"The line at the bottom?" I asked, pointing to the picture.

"Well," he said, "in the context of star gazing and constellations, it would seem logical to assume the line represents the horizon."

I had to smile. "Of course!" I exclaimed, slapping my head. "Why not? That's perfect."

"Yes," he responded, "I thought it was pretty good myself. Now the intriguing thing about the appearance of the Wintermaker in the night sky, coming in all his majesty over the horizon, was recognized by Robert Frost when he wrote:

> You know Orion always comes up sideways,
> Throwing a leg up over our fence of mountains
> And rising on his hands, he looks in on me
> Busy outdoors . . .

"This, to me, was the answer to the eastern orientation of the figures at Hegman Lake; the Wintermaker makes a dramatic leap into the heavens in early December. He lifts clear of the horizon in the evening of December 21, the winter solstice.

"Now, Thaddeus Cowen has suggested the effigy mounds in the Ohio Valley were built to represent star patterns. It would be as if the sky spirits could look down and see their images and relate to the people who built them. At Hegman Lake, in that same context, the Wintermaker leaps into the cold sky, and sees his shining image on a frozen cliff.

"This idea brought home to me a basic fact: the Ojibwe were not farmers. Farmers, like the Anasazi, the Maya, the early Britons, all created solar calendars. So did the Indians of the northern plains, the ones who built the medicine wheels that were oriented to the solstice sunrise. The people of Hegman Lake were hunters. They welcomed the Wintermaker, and celebrated his coming with a snowshoe dance.

"The Ojibwe did this because during the long winter months, hunters could cross frozen lakes and swamps and overtake caribou, deer, and moose. Meat could be preserved easily on raised platforms, and fur bearers came into prime. Winter to these hunters did not mean death and decay, but opportunity and sustainability.

"If you ever take a look at the topographic maps of the Hegman Lake area, it is easy to surmise the Hegman Lake hunters wintered there. They may have spent their summers on Lac La Croix, Crooked, or Basswood Lake, where they fished, picked blueberries, and harvested wild rice. At freeze-up, they could easily walk inland, following what is known as the Spring Creek Draw, and crossing the ridges eastward to Hegman Lake, the Wintermaker's own shrine.

"The topographic map also tells another story. The granitic rocks with the pictographs have a southern as well as eastern orientation. Look down the lake—no, the other way—see the high hill, the open dome of granite? The significance of this is that on the spring equinox, and at least two weeks before it, the Wintermaker stands above that knob to mark the end of winter. Thus this site actually has two calendar orientations, and each one is focused on and around the Wintermaker.

With snoeshows, hunters could cross frozen lakes and swamps and overtake caribou, deer, and moose.

"Last, but certainly not least, comes the final clue provided by my father and grandfather. It was hidden in their story of the great primordial flood caused by Nanaboujou, who defeated Great Panther through trickery, and by doing so lost his world to water.

"And you now know what I finally realized. When the constellation of the great panther rises in the spring sky, it is to say so-long to the ice, and hello to meltwater and floods. This may be the practical reason for the stellar calendar of these hunters: when the panther reached its ascendancy, one must get out of winter camp as fast as possible, for the panther brings with it the promise of spring floods.

"As for the other markings . . . well," he slowly said, "it has been suggested that the six horizontal lines represent the number of lunar months that the Wintermaker can be seen in the evening sky, and the three stars in a vertical line may be the three stars in the Wintermaker's knife.

"That leaves us with the short stroke of paint above the moose's head. If the pictures are constellations, then perhaps the hashmark represents a comet, or possibly a shooting star. If it was a comet, then that means it was a comet that appeared in Pegasus during the midwinter. If astronomers could find such a comet, then perhaps the paintings could be dated."

"So what you're saying is that right above us we have a time marker, almanac, and star chart all rolled into one."

"Absolutely!" he cried. "A wonderful device to remind the hunters to make haste across the ridges and frozen waterways before the spring meltdown led to floods and impassable ground. When the Wintermaker touched the bare, granite dome it was time to make haste to the sugarbush camp for the happy reunions with other wintering families, and there remember the loved ones who had passed away, and to once again prepare for the sweet maple sugar harvest.

"The Hegman Lake artist created right here a mnemonic bible of the Indian spirit world that was reflected in the sky. On this site, the entire universe turned slowly with the seasons, reaffirming the place of Native Americans in the cosmos. Listen to this, written by Maude Kegg," and he recited by memory:

> *It's long ago and the Indians are cold because the*
> *winter is too long. They make bows for the children.*
> *They tell them: Go and shoot up in the sky. Shoot the*
> *Wintermaker. The children go outside and aim*
> *skywards. They shoot the Wintermaker. And sure*
> *enough it warms up.*

"Got an extra bow and arrow?" I asked. "I would like to see how long this warm-up takes."

Earth Walks laughed. "So," he demanded, "what do you think of it all?"

"I'm impressed all to hell," I answered. "It makes sense, and the amount and quality of the research you've done is pretty amazing."

"Thank you. I appreciate that. But remember, it was my grandfather who started this journey, and my father who continued it. Both of them believed in the Ojibwe sky and the traditions and stories that can be seen overhead on any clear evening. Today, when I talk about rock paintings or constellations, I try and also tell the audience one or two of my father's sky stories and the traditions and myths they represent. I also tell them that the people who painted these pictures lived in a much different world than we do, but in the end they were:

> *Dream catchers,*
> *Star gazers,*
> *Earth watchers,*
> *Myth makers,*

"And, even in this day and age, these are not bad things to be."

"Not bad things at all," I agreed, wondering if he would include me in any of those categories.

CHAPTER NINE

THE NEVER-ENDING CIRCLE

"Damn habit forming," Earth Walks said.

Looking up from my notes, I stared at him, not knowing what he meant or where he was coming from. We had been meeting twice a week for the past two months in preparation for the upcoming summer program. Just now we had been discussing the possibility of a two- or three-day bus tour, either south to Jeffers, Pipestone, and the Minnesota River valley, or north to Grand Portage, the North Shore volcanics, and Lake Superior. I wanted to go north, Earth Walks south.

"Habit forming you know," Earth Walks repeated, louder than before.

"Habit forming?" I echoed, my thoughts on field excursions evaporating like water in a desert lake. "What on earth are you talking about?"

"That's exactly it," he cried, opening his arms to embrace the entire landscape. "Earth! Our planet! This world! We grow up, grow older, it all becomes like a sunrise, spring rain, morning cup of coffee—something always there, ho-hum, a habit. The wonder, awe, mystery—banished to the Disney channel.

"You know," he went on, "when you first came up to my little camp and told me how magnificent and beautiful the trees were, you reopened my eyes. I had stopped seeing those great star catchers as anything other then a part of the everyday landscape, the usual, but to go, please, for I'm in a hurry. Me! Of all people—can you believe it?"

"It happens to all of us," I said. "To those of the spirit as well as those of the rock hammer. I lived in Vancouver, British Columbia, for a couple of years and had this really neat apartment. It was on the fourteenth floor of a high-rise and had an incredible view of Stanley Park, the Coast Range, and the harbor. When I first moved in I spent long afternoons and evenings staring out the picture windows or sitting on the balcony, lost in the beauty and magnificence, thinking how lucky I was to be alive and part of something so awesome. But too soon that view became just another picture hanging on my wall. I knew it was there, constantly saw it, stopped and looked at it once in a while, but it was no longer the same as when I first opened those curtains—it had become commonplace, or, as you say, the usual."

"Like our lives," he sadly offered, shaking his head. "We all too soon lose our wonder, appreciation, and zest. Life too becomes commonplace, a habit, just more of the usual."

"Ah, sweet mystery of life," I sang in my usual off-key voice. "Where art thou gone? North or south? Do we care? Does it matter? And that's the end of this dreadful song!"

"All right," he howled, holding up his hands. "You sound worse than my old washing machine, and that got so noisy I dragged it to the dump and shot it! And yes, it does matter and I do care, as you darn well know! Not to be awestruck, amazed, inquisitive, questioning, mystified, exalted, not to get goosebumps, chills, feelings of reverence and respect for this world we inhabit and the miracle of genetic evolution that is us, means the spirit and mind are not much different than a bowl of grape Jell-O sitting in the summer sun.

And," he continued, "this is where I think that not only are you and I, science and the Native American holistic view of this land and planet, compatible, but each is actually necessary to ensure the health and well-being of the other. To know that continents move, that Mount Everest is so high because India is colliding with Asia, that there are strange life-forms down in the dark, on the bottom of the sea floor, that thrive without sunlight, to be able to understand earthquakes, predict and witness volcanic eruptions, decipher glacial deposits, watch hummingbirds dance, see a wolf pack encircle a starving moose, watch an Indian sit on the ground listening to Mother Earth, dance in a powwow, see sun rise on the winter or spring equinox, or be like a laughing child and run into a meadow full of wild flowers and vanish from reality for a short while—goosebumps, elation, humility, reverence, respect, and imagination. This is part spirit, part science—a moving, never-ending circle of stars, animals, geology, land, and us, all connected and all beginning and ending with planet earth.

"Thanks to you, over the last few months I have come to see and appreciate that geological studies of Mother Earth reveal her nature in fascinating wonder and great detail. Though the science of geology is relatively young, it has brought a sense of order to the things Native Americans knew, sensed, and believed, but could not explain in such a systematic way. It has also shown us natural processes we were not aware of or had no understanding of.

"Even so, I don't think there is much doubt my ancestors were geologists, as well as astronomers, botanists, ecologists, chemists, and medical doctors. Think about what they knew and did. They never lived on floodplains, they knew how to use or avoid over two hundred species of plants, as well as how and when they all grew, they knew how to mine and use many different kinds of rocks and minerals, they knew about pain killers, birth control, and hallucinogenic drugs before Europeans set foot on this continent.

"They knew that North America was surrounded by water, that melting ice caused great valleys and large floods, they knew the shape of Lake Superior, the summer, spring, and winter equinoxes, they figured out the days of the lunar month, and had an orderly vision of the sky that equaled that of the Greeks and Romans.

"But they were not supermen, saints, or Johnny Appleseeds, they were people who had to know all of this and much more just to survive. Without this knowledge, if they had remained ignorant savages, they would soon have followed in the path of the mastodon and giant sloth.

"Geology and the culture of my people, two seemingly separate ways, yet together I believe they can show us how to live with and care for the land, how to restore and maintain the balance among all living things, physical and spiritual."

As he spoke, his words echoed from the cedar walls, they rolled across the room like summer thunder, they found and embraced me like a cold rain.

Finally he fell silent, and all I could think to do was to refill my glass with some of the twenty-one-year-old Bowmore Scotch I had given him. I knew he didn't drink, but then I also knew it would be here when I came to visit. Swallowing a mouthful, I set the glass down and took a deep breath.

I said, "I am rather pleased you see geology in this way, for I know that wasn't always the case. Over the past few months," I continued, taking another sip of the smooth, smoky liquid, "I have come to see that science has multiple roles to play in our society and our lives. If we skip the most obvious—to make life safer, longer, healthier, more comfortable for us all—I think the first and foremost of these is to find and explain the unbiased truth. And to do so in its own unique way of thinking—what is called the scientific method. Second, science expands our horizons, pushes us to our limits and beyond. Finally, it attempts to understand and explain planet earth: how our planet works, what it is made of, the delicate balance between life, water, atmosphere, the moving crust, and our part and place in all this.

"On the other hand, as I have said before, there are many mysteries and happenings that science is not yet able to explain, many questions science has no answer for, or no proof one way or the other. I imagine this as the doorway to your spiritualism.

"You once told me my scientific view of planet earth had no balance, that I saw with one and only one eye, and that the Western philosophy of conquest and exploitation of nature and planet earth was evil and morally wrong."

Earth Walks held up his hands as if to say something, but, wanting to finish before I lost my train of thought, I rushed on.

"I don't necessarily agree with you on all aspects of this," I said. "But I have come to believe you are right about one important thing—it would be much better and healthier for us and for planet earth if we all saw, or tried to see, with two eyes instead of one. I believe your holistic approach to planet earth, your kind of spiritualism, can help us do this, help keep us on an even keel, give us balance, keep us human, humble, respectful, feeling, and, most important, laughing and smiling.

"And, speaking of balance and humanity, I also believe it is the responsibility of your kind of spirituality to help protect us from the dark side of science and those who practice it. Science, as a profession, has done a poor job of policing itself, of enforcing or even having a code of ethics or, sometimes, even a sense of moral obligation. There are scientists for whom the experiment, end result, data collection, next grant, next publication, power, or money is all that matters. These are the end-justifies-the-means crowd, the ones who brought us, at incredible expense to the taxpayer, nuclear weapons, radiation experiments, acid rain, Agent Orange, CFCs, ozone holes, polluted air and water, nuclear meltdowns, nuclear and chemical waste, oil spills, clear-cut valleys, and on and on long into the night.

"As Martin Luther King said, 'The means by which we live have outdistanced the ends for which we live. Our scientific power has outrun our spiritual power. We have guided missiles and misguided men.' It's up to people like you to get back in the race, to use your 'spirit power' to turn the lights on in the laboratories where this kind of scientist works as well as sometimes hides. You need to keep constant watch outside their doors to ensure these supposedly intelligent and unbiased human beings consider and pay attention to the long-range consequences of what they do. To help them care for, be responsible to, and respect planet earth and the life that lives upon it. The alternative might turn out to be extremely expensive for us all—a world without humans, or, in a technological sense, a world without humanity. And now," I said with a pause, "just where were we?"

"On common ground," Earth Walks replied, giving me a pleased look. "And that is somewhere along a never-ending circle that is geology, Native Americans, modern society, planet earth, and the two of us!"

BIBLIOGRAPHY

Allaby, Michael. *A Guide to Gaia.* New York: Dutton, 1989.

Allman, William, F. "The Dawn of Creativity." *U. S. News and World Report,* May 1996.

Ball, Timothy. "Climatic Change, Droughts and Their Social Impact: Central Canada, 1811–1820, A Classic Example." In *The Year without a Summer? World Climate in 1816,* edited by C. R. Hurington. Ottawa: Canadian Museum of Nature, 1992.

———. "The Year without a Summer: Its Impact on the Fur Trade and History of Western Canada." In *The Year without a Summer? World Climate in 1816,* edited by C. R. Hurington. Ottawa: Canadian Museum of Nature, 1992.

Begley, J. and A. Murr. "The First Americans." *Newsweek,* April 1999.

Bishop, C. A. *The Northern Ojibwa and the Fur Trade, An Historical and Ecological Study.* Culture and Community Series, edited by S. M. Weaver. Toronto: Holt, Rinehart, and Winston, 1974.

Bleger, Theodore C. *Building Minnesota.* Boston: D. C. Heath and Company, 1938.

Bolz, Arnold, J. *Portage into the Past, By Canoe along the Minnesota-Ontario Boundary Waters.* Minneapolis: University of Minnesota Press, 1960.

Bonnichsen, Robson, Dennis Stanford, and James Fastook. "Environmental Change and Developmental History of Human Adaptive Patterns: The Paleoindian Case." *Decade of North American Geology, North America and Adjacent Oceans During the Last Deglaciation,* vol. K-3, edited by W. F. Ruddman and H. E. Wright. Boulder, CO: Geological Society of America, 1987.

Caduto, Michael, and Joseph Bruchac. *Keepers of the Earth.* Golden, CO: Fulcrum, 1988.

Campbell, Margorie. *The North West Company.* Toronto: MacMillian Co. of Canada, 1973.

Canby, Thomas Y. "The Search for the First Americans." *National Geographic,* Sept. 1979.

Catlin, George. *Letters and Notes on the Manners, Customs, and Conditions of the*

North American Indians, vols. 1 and 2. New York: Dover Publications, 1973.

Circle Trail. Pipestone: Pipestone Indian Shrine Association, 1959.

Conway, Thor. *Painter Dreams, Native American Rock Art.* Minocqua, Wis.: Northwood Press, 1993.

Danziger, Edmund, J. *The Chippewas of Lake Superior.* Norman: University of Oklahoma Press, 1978.

Deloria, Vine, Jr. "The Bering Strait and Narrow." *Winds of Change,* Winter 1995.

Densmore, Francis. *Chippewa Customs.* Minneapolis: Ross and Haines, 1970.

———. *How Indians Use Wild Plants as Food, Medicine, and Crafts.* New York: Dover Publications.

Dewdney, Selwyn. *Indian Rock Painting of the Great Lakes.* Toronto: University of Toronto Press, 1967.

Doll, Don. *Vision Quest.* New York: Croen Publishing, 1994.

Drier, Roy, and Octave DuTemple. *Prehistoric Copper Mining in the Lake Superior Region.* Published privately, 1961.

Eastman, Charles, and Elaine Goodale Eastman. *Wigwam Evenings.* Lincoln: University of Nebraska Press, 1990.

Eddy, John A. "Astronomical Alignment of the Big Horn Medicine Wheel." *Science* 184 (1974).

———. "Medicine Wheels and Plains Indian Astronomy." In *Astronomy of the Ancients,* edited by Kenneth Brecker and Michael Feirtag. Cambridge: MIT Press, 1981.

Erdoes, Richard, and Alfonso Ortiz. *American Indian Myths and Legends.* New York: Pantheon Books, 1984.

Ericson, Jonathon, and Barbara Purdy. *Prehistoric Quarries and Lithic Production.* New York: Cambridge University Press, 1984.

Fagan, Brian M. *The Great Journey.* New York: Thames and Hudson, 1987.

Fitting, James. "Environmental Potential and the Postglacial Readaptation in Eastern North America." *American Antiquity* vol. 33, no. 4 (1968).

Fletcher, Alice. "The Hako: A Pawnee Ceremony." *Smithsonian Institution of American Ethnology Annual Report* 22, part 2 (1904).

Fredel, Stuart J. *Prehistory of the Americas.* Cambridge: Cambridge University Press, 1987.

Fredrickson, N. Jaye. *Indian Ceremonial and Trade Silver.* Ottawa: National Museum of Canada, 1980.

Frost, Robert. *Collected Poems, Prose, and Plays.* New York: The Library of America, 1995.

Gilman, Carolyn. *The Grand Portage Story.* St. Paul: Minnesota Historical Society, 1992.

Goodman, Jeffery. *American Genesis.* New York: Summit Books, 1981.

Grant, Cambell. *Rock Art of the American Indian.* New York: Thomas Y. Crowell Company, 1967.

———. *The Rock Art of the North American Indians.* New York: Cambridge University Press, 1983.

Griffen, James B., ed. *Lake Superior Copper and the Indians: Miscellaneous Studies of Great Lakes Prehistory.* Ann Arbor: University of Michigan Press, 1961.

Grove, Noel. "The Superior Way of Life." *National Geographic,* December 1993.

Harmon, Daniel Williams. *Sixteen Years in the Indian Country: The Journal of Daniel William Harmon, 1800–1816.* Toronto: MacMillian Co. of Canada, 1957.

Harvey, Karen D. *Indian Country.* Denver: North American Press, 1994.

Hungry Wolf, Beverly. *The Ways of My Grandmothers.* New York: Quill, 1982.

Huyghe, Patrick. *Columbus Was Last.* New York: Hyperion, 1992.

Innis, Harold A. *The Fur Trade in Canada.* Toronto: University of Toronto Press, 1970.

Jelinek, Arthur. "Man's Role in the Extinction of Pleistocene Faunas." In *Pleistocene Extinctions, The Search for a Cause.* New Haven: Yale University Press, 1967.

Johnson, Eldon. *The Prehistoric Peoples of Minnesota,* 3rd ed., St. Paul: Minnesota Historical Society Press, 1988.

Johnston, Basil. *By Canoe and Moccasin: Somenative Place Names of the Great Lakes.* Lakefield, ON: Waapoone Publishing, 1986.

———. *Ojibwe Heritage.* Toronto: McCelland and Stewart, 1990.

Kegg, Maude. *Portage Lake.* Minneapolis: University of Minnesota Press, 1993.

Keyser, James, and Linea Sundstrom. *Rock Art of Western South Dakota.* Vermillion, SD: South Dakota Archaeological Society, 1984.

Kohl, George Johann. *Kitchi-Gami.* St. Paul: Minnesota Historical Society Press, 1985.

Krupp, E. C. "A Glance into the Smoking Mirror." In *Archaeoastronomy in the Americas,* edited by Ray Williamson. Altos, Calif.: Ballena Press, 1981.

Kupferer, Harriet. *Ancient Drums, Other Moccasins: Native North American Cultural Adaptation.* New York: Prentice Hall, 1988.

Kwsey, Howard. *Strange Empire: A Narrative of the Northwest.* New York: Morrow, 1952.

Laberge, Gene L. *Geology of the Lake Superior Region.* Phoenix: Geoscience Press, 1994.

Lemonick, Michael. "Coming to America." *Time,* May 1993.

Levenson, Thomas. *Ice Time.* New York: Harper and Row, 1989.

Levi, Carolissa. *Chippewa Indians of Yesterday and Today.* New York: Pageant Press, 1956.

Longfellow, Henry Wordsworth. *Hiawatha.* Lancaster: Jacques Cattell Press, 1944.

Lothson, Allen Gordon. *The Jeffers Petroglyphs Site.* St. Paul: Minnesota Historical Society Press, 1976.

Lovelock, J. E. *Gaia: A New Look at Life on Earth.* Oxford: Oxford University Press, 1979.

Lydecker, Ryck. *The Edge of the Arrowhead.* Minneapolis: Minnesota Marine Advisory Services, Office of Sea Grant, 1976.

Mallery, Garrick. *Picture Writing of the American Indians,* vol I. New York: Dover Publications, 1972.

Martineau, LaVan. *The Rocks Begin to Speak.* Las Vegas: KC Publications, 1973.

Mason, Carol. *Introduction to Wisconsin Indians.* Salem, Wis.: Sheffeld Publishing, 1988.

Mason, Ronald J. *Great Lakes Archaeology.* New York: Academic Press, 1981.

Matsch, Charles. "River Warren, the Southern Outlet of Glacial Lake Agassiz." In *Glacial Lake Agassiz,* edited by James Teller and Lee Clayton, pp. 231–244. Geological Association of Canada Special Paper 26. St. John's: Memorial University, 1983.

McCracken, Harold. *George Catlin and the Old Frontier.* New York, Bonanza Books, 1969.

McLuhan, T.C. *Touch the Earth.* New York: Simon & Schuster, 1971.

Miles, Tom and George Hoffman, dirs. *The Pipe Makers.* Woodpecker Productions, 1994. Video recording, 38 mins.

Minnesota Chippewa Tribe. *Kitchi Onigaming.* Cass Lake: Minnesota Chippewa Tribe, 1983.

Morton, Ron L. *Music of the Earth.* New York: Plenum Press, 1996.

Northwest Interpretive Association. *The Great Floods, Cataclysms of the Ice Ages, Grand Coulee Dam.* Washington. Video recording.

Nute, Grace Lee. *Lake Superior.* New York: Bobbs-Merrill Co., 1944.

Ojakangas, Richard W., and Charles L. Matsch. *Minnesota's Geology.* Minneapolis: University of Minnesota Press, 1982.

Petit, C. W. "Rediscovering America." *U.S. News & World Report,* October 1998.

Pettipas, Leo. F., and Anthony P. Buchner. "Paleo-Indian Prehistory of the Glacial Lake Agassiz Region in Southern Manitoba, 11500 to 6500." In *Glacial Lake Agassiz,* edited by James Teller and Lee Clayton, pp. 421–449. Geological Association of Canada Special Paper 26. St. John's: Memorial University, 1983.

Pielou, E. C. *After the Ice Age.* Chicago: University of Chicago Press, 1991.

Quimby, George I. *Indian Life in the Upper Great Lakes, 11000 B.C. to A.D. 1800.*

Chicago: University of Chicago Press, 1960.

———. *Indian Culture and European Trade Goods.* Madison: University of Wisconsin Press, 1966.

Rapp, George, Jr., E. Henrickson, and James Allert. "Native Copper Sources of Artifact Copper in Pre-Columbian North America." In *Archaeological Geology of North America,* edited by N. P. Lasca and J. Donahue. Centennial Special Volume 4. Boulder, Colo.: Geological Society of America, 1990.

Rapp, George, Jr., and C. Hill. *Geoarchaeology.* New Haven: Yale University Press, 1998.

Readers Digest Association. *Mysteries of the Ancient Americas.* Pleasantville, NY: Readers Digest, 1986.

Rhodes, Frank H. T., and Richard O. Stone. *Language of the Earth.* New York: Pergamon Press, 1981.

Rogers, John. *The Red World and White.* Norman: University of Oklahoma Press, 1957.

Ross, Thomas E., and Tyrel G. Moore. "Indians in North America." In *A Cultural Geography of North American Indians,* edited by Thomas Ross and Tyrel Moore. Boulder, CO: Westview Press, 1987.

Rothman, Hal, and Daniel Holder. *Managing the Sacred and the Secular: An Administrative History of Pipestone National Monument.* Henderson, NV: National Park Service, Midwest Region, 1992.

Roufs, Timothy. *Anishinabe of the Minnesota Chippewa Tribe.* Phoenix, AZ: Indian Tribal Series, 1975.

Roufs, Timothy, and Larry Aitken, eds. "Information Relating to Chippewa Peoples." In *Handbook of American Indians North of Mexico.* Duluth, Minn.: Lake Superior Basin Studies Center, 1984.

Satz, Ronald N. "Chippewa Treaty Rights," *Transactions,* vol. 79. Madison: Wisconsin Academy of Science, Arts, and Letters, 1991.

Severance, Cordenio. *Indian Legends of Minnesota.* New York: D. D. Merritt Company, 1893.

Shaffer, Lynda Noeme. *Native Americans before 1492.* Armonk, NY: M. E. Sharp, 1992.

Sigurdsson, Haraldur, and Steven Carey. "The Eruption of Tambora in 1815: Environmental Effects and Eruption Dynamics." In *The Year without a Summer? World Climate in 1816,* edited by C. R. Hurington. Ottawa: Canadian Museum of Nature, 1992.

Silver, Cheryl Simon. *One Earth, One Future.* Washington: National Academy Press, 1990.

Sofaer, Anna. *Sun Dagger, Solstice Project.* Bullfrog Films, 1983. Video recording.

Sofaer, Anna, Zinser Volkar, and Rolf Sinclair. "A Unique Solar Marking Construct." *Science* 206 (1979).

Standing Bear, Luther. *Land of the Spotted Eagle.* Boston: Houghton Mifflin, 1933.

———. *My People the Sioux,* edited by E. A. Brininstool. Boston: Houghton Mifflin Company, 1928.

Steinbring, John. "Copper Technology during the Archaic Tradition." Ph.D. thesis, University of Minnesota, 1975.

Strand, Dale. "Pipestone Quarry, Then and Now." Paper for history 8-893, University of Minnesota-Duluth, 1973.

Strommel, Henry, and Elizabeth Strommel. *Volcano Weather.* Newport: Seven Seas Press, 1983.

Suffling, Roger, and Ron Franks. "The Ecology of Famine: Northwestern Ontario in 1815–1817." In *The Year without a Summer? World Climate in 1816,* edited by C. R. Hurington. Ottawa: Canadian Museum of Nature, 1992.

Tanner, Helen Hornbeck. *The Ojibwa.* New York: Chelsea House Publishing, 1992.

Vastokas, R. "Aboriginal Use of Copper in the Great Lakes Area." Ph.D. thesis, Columbia University, 1970.

Vennum, Thomas, Jr. *Wild Rice and the Ojibwe People.* St. Paul: Minnesota Historical Society Press, 1988.

Vitaliano, Dorothy. *Legends of the Earth.* Secaucus: Citadel Press, 1973.

Warren, William. *History of the Ojibway People.* St. Paul: Minnesota Historical Society Press, 1984.

Watthall, John A. *Galena and Aboriginal Trade in Eastern North America.* Scientific Papers, vol. 17. Springfield: Illinois State Museum, 1981.

Willey, Gordon R. *An Introduction to American Archaeology,* vol. 1. Englewood Cliffs: Prentice Hall, 1966.

Williamson, Ray A. *Living the Sky.* Boston: Houghton Mifflin Company, 1984.

Woolworth, Alan. *The Red Pipestone Quarry of Minnesota: Archaeological and Historical Reports.* St. Paul, MN: Minnesota Archaeological Society, 42, (1924).

Wright, H. E., Jr. *Geologic History of Minnesota Rivers.* Educational Series 7. St. Paul: Minnesota Geological Survey, 1990.

———. "The Environment of Early Man in the Great Lakes Region." In *Aspects of Upper Great Lakes Anthropology,* edited by Eldon Johnson. St. Paul: Minnesota Historical Society Press, 1974.

Young Bear, Severt. *Standing in the Light: A Lakota Way of Seeing.* Lincoln: University of Nebraska Press, 1994.

INDEX